BASIC FLY TYING

Ed Koch and Norm Shires

• Stackpole Books •

Copyright © 1990 by Stackpole Books

Published by
STACKPOLE BOOKS
5067 Ritter Road
Mechanicsburg, PA 17055

All rights reserved, including the right to reproduce this book or portions thereof in any form or by any means, electronic or mechanical, including photocopying, recording, or by any information storage and retrieval system, without permission in writing from the publisher. All inquiries should be addressed to Stackpole Books, 5067 Ritter Road, Mechanicsburg, Pennsylvania 17055.

Printed in the United States of America

10 9 8 7 6 5 4

First Edition

Cover and interior photographs by Norm Shires
Cover design by Tracy Patterson
Interior design and typesetting by Art Unlimited

Library of Congress Cataloging-in-Publication Data

Koch, Ed.
  Basic fly tying / Ed Koch and Norm Shires. — 1st ed.
    p. cm.
  ISBN 0-8117-2318-6
  1. Fly tying. I. Title.
SH451.K588 1990
688.7'912—dc20
                      90-31613
                      CIP

Contents

Introduction: How to Use This Book   5
Part One: Tools and Materials   7
Part Two: Basic Techniques   **17**
  Threading the Bobbin   19
  Placing the Hook in the Vise   19
  Tying On Thread and Tying the Half-Hitch   20
  Tying the Whip-Finish   24
  Weighting a Hook   26
  Applying Head Cement   29
Part Three: Tails   **31**
  Hackle-Fiber Tail   32
  Deer-Hair Tail   38
  Split Tail   42
Part Four: Bodies   **47**
  Dubbed Fur Body   48
  Dubbed Fur Body with Tinsel Rib   52
  Chenille Body   58
  Peacock Herl Body   62
  Clipped Deer-Hair Body   66
  Quill Body   74
  Thorax and Wing Case   78
  Foam Body   84
Part Five: Wings   **89**
  Upright Wood-Duck Wings   91
  Poly Wings, Spent   96
  Hackle-Tip Wings   100
  Tent-Type Wings   104
Part Six: Hackle   **111**
  Winding Dry-Fly Hackle   113
  Wet-Fly Hackle   116
  Palmered Hackle   122
Afterword   **126**
Appendix   **127**

*To Betty Ann and Ruth,
our patient and understanding wives*

---

*And to
all beginning fly tiers everywhere*

Introduction

How to Use This Book

Basic Fly Tying is just what the title says—a book about the most basic techniques in the craft of tying flies. Every discipline, whether it's oil painting, auto mechanics, or golf, builds on fundamental methods and skills. So it is with fly tying. We've gathered together in this book what we believe are the essential techniques upon which further development in tying flies is based. Master these techniques and you can, with patience and practice, master any of the more difficult operations encountered in tying flies.

We've divided the book into sections covering the fundamental techniques used in tying any fly, techniques for different tails, wings, and bodies, and a section on winding hackle for dry and wet flies. Only a few of the exercises in these sections result in a complete fly you can fish with, and this was something we intended. Rather than teach you highly specific techniques that may be used to tie only a few flies, we've focused on those more universal techniques that are used in virtually all fly patterns. This is a book of skills, not of fly recipes. When you've worked through and mastered the various separate operations covered in this book you will be able to follow the directions in almost any fly pattern successfully.

We recommend you proceed through this book at the pace you're most comfortable with, taking less time on those operations you find easier and more time on those you find difficult. Tying flies is a manual craft and carries with it all the frustrations and joys to be had when one person sits down and creates something with his or her hands. You're going to be exasperated at times: you'll break thread at critical moments, you'll drop materials just when you thought you had secured them on the hook, your tails or wings or hackle won't center properly and you'll have to scrap the materials you're working on. But these are things *all* fly tiers have experienced at one time or another and they can be instructive; sometimes you can only get it right after you recognize what it means to get it wrong. Stick with it, practice, and you'll find your efforts richly rewarded.

The instructions in the tying exercises to follow are given with the assumption that the tier is right-handed. Some left-handed tiers have learned to tie right-handed and you may want to give this a try. Both hands are used almost equally in fly tying—it cannot be said to be a craft in which right-handers have the advantage. It may help you to mount the vise opposite the direction a right-hander would find comfortable; if you do this, take care when following fly patterns to reverse the direction of your wraps.

Each exercise to come will describe the specific materials you will need to follow it. When you begin to tie flies from new or established patterns, you will follow the pattern's requirements for hooks and materials.

There are many, many variations in tying techniques, and we believe the ones we have included in this book are the easiest for the beginning tier to master quickly. It is one of our goals in preparing this book to give you a firm basis of fundamental skills from which you can branch out on your own to experiment, learn new techniques, and grow as a fly tier. We hope this primer serves you well as you begin your journey into the creative, many-layered, and rewarding world of fly tying.

Tools and Materials

Part 1

- Vise
- Scissors
- Bobbin
- Bodkin
- Bobbin Threader
- Hooks
- Half-hitch Tool
- Necks
- Thread
- Head Cement
- Feathers
- Hair
- Dubbing Fur
- Synthetic Materials

The tools and materials described in this section are what we feel to be the basic necessities for the beginning tier. There are many more tools, gadgets, and highly specific materials on the market, and as you become more proficient you will want to explore the wealth of fly-tying products available. But for now, concentrate on developing your skills with these fundamental tools and materials—they are the foundation from which all fly tiers begin.

Getting started need not be an overly expensive proposition. It is possible to outfit yourself with the basic tools described in this section for under a hundred dollars. Having said that, though, we recommend you buy the best tools you can afford. Fortunately, "best" doesn't have to mean "most expensive." By doing a little comparison shopping, studying catalogs, visiting fly shops, and talking to experienced tiers, you can discover which tools are of the highest quality in your price range. Begin by requesting some of the catalogs for the companies listed in the Appendix. Enrolling in a beginning fly-tying class is another good way to find help in selecting your first set of tools and materials.

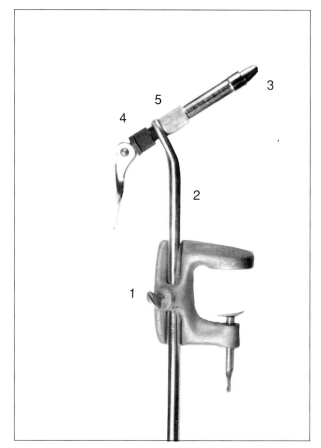

The Vise

The Vise

The function of a vise is to hold the hook securely as the tier applies various materials to it to fashion a fly. Prices vary widely, but here are some features the beginning tier should look for in purchasing a first vise:
- a durable mounting mechanism or base
- a jaw that will hold a range of hook sizes securely
- a design that lets you work around the hook easily
- components that won't rust or corrode
- easy and convenient adjusting of jaw tension and rotation

Here's another thing to keep in mind: A highly polished and shiny vise may bother your eyes after a time; consider a vise made of burnished or darker metal.

The photograph illustrates the basic components of a fly-tying vise.
1. C-clamp for fastening the vise to the tying table or bench. Most C-clamps will open to a minimum of one and a half to two inches, making the vise usable on any surface from a three-quarter-inch-thick tying table to a one-and-three-quarter-inch-thick picnic table in a campground. Some vise models do not use C-clamps, but are mounted on a secure, heavy base that sits atop the tying table. Both mounting designs work well; choose whichever appeals to you.
2. The rod, incorporating on one end the C-clamp, and fastened on the other end to a collar secured to the jaws, which grip the hook.
3. The jaws.
4. The knob or wheel, used to adjust jaw tension.
5. A wing nut that, when loosened, permits the collar and jaws to rotate in a complete circle. Most vises today allow jaw rotation, although the mechanisms permitting the rotation vary.

The D. H. Thompson Company (see Appendix), one of the oldest manufacturers of fly-tying equipment in the United States, still makes one of the best and least-expensive vises on the market today—the Thompson Model A, which is shown in the photo. We highly recommend it for beginning tiers on a limited budget. The Universal Vise Company (see Appendix) is the originator of the rotating-jaw vise. They market two models today that we also recommend to the tier in search of his or her first vise.

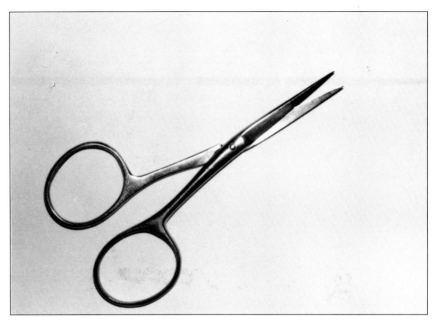

Scissors

Scissors

Good scissors are an essential tool to the fly tier. They are used to cut and fashion various materials: thread, feathers, animal hair, synthetic materials. You should start out with two pairs: one three to three-and-a-half inches long with fine tips for precision work on small flies; and a larger pair five to five-and-a-half inches long for heavy work (cutting deer hair and synthetics).

Look for these features in selecting your first pairs of scissors:
- finely tapered, sharp points that come together evenly
- blades that mesh evenly and shear cleanly (serrated edges are a plus in cutting heavy materials)
- finger loops that accommodate the thumb and fingers comfortably
- stainless steel or tungsten steel construction
- a total length that will give you enough clearance to easily see past your hand to the fly you are working on

There are many features available in tying scissors. Some have mag-

Tools and Materials

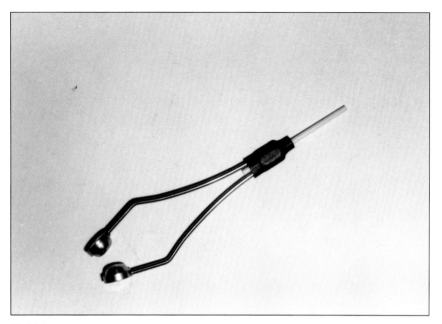

Bobbin

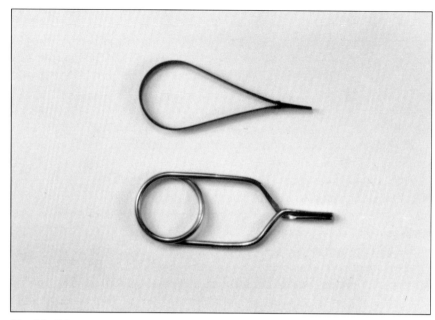

Hackle Pliers

netized tips for picking up small hooks, some have curved blades, some have serrated blades, the list goes on. Try not to be overcome by the options available as you learn to tie. Keep it simple at first, concentrate on the points outlined above, and be sure, if you are left-handed, to buy scissors made for you—they do exist and are a must if you are a lefty.

It is possible to buy a good-quality pair of scissors for under twenty dollars. The D. H. Thompson Company offers a good, reasonably priced selection, as does Umpqua Feather Merchants (see Appendix).

Bobbin

The bobbin is a tool that helps conserve and control tying thread by clipping onto a spool and dispensing the thread at constant tension. The tier manipulates the bobbin, instead of the spool and thread itself, and because the thread comes out of the bobbin at an even tension, it is easier for the tier to apply the proper amount of pressure when wrapping thread. And with a bobbin the tier can allow the thread to hang freely from the hook while he or she attends to matters requiring two hands. Try doing that with a spool of thread!

There are several bobbin designs available, but the wishbone-type pictured is the most popular. The two arms clip onto the spool and the thread is pulled from the spool and passed through the bobbin tube. Tension on the spool is adjusted by gently flexing the bobbin arms; thus, lighter or heavier tension can be achieved.

As simple a device as the bobbin is, there are yet things to look for when buying one:
• rigid arms that will not twist
• enough weight to maintain adequate pressure when the thread is wrapped around the hook and left to dangle as other operations are performed
• a thread tube that is smooth (no burrs or nicks) and of a diameter small enough to allow control but not so small as to scrape the thread
• the ability to accommodate a range of spool sizes

Bobbins are inexpensive tools, and a well-made one will last for years. You may wish to have more than one; that way, the thread you use most can have its own bobbin that is threaded up and ready to go any time you are.

Hackle Pliers

Hackle pliers are used during certain tying operations to grip small, hard-to-handle materials. They are a must when hackling dry flies.

We recommend two sizes: a small pair for dry flies of #16 or smaller, and a larger pair for flies #14 and larger. Look for these features:
- well-adjusted jaws that will grip firmly without cutting
- enough weight to apply adequate tension when clipped to a particular material and left dangling
- easy and comfortable to grip and easy to open

Hackle pliers are inexpensive. The D. H. Thompson Company makes an excellent pair at a reasonable price, as do most other tying-tools companies. Two styles are shown in the photo.

Half-Hitch Tool

The half-hitch is a knot used frequently in fly tying to secure thread wraps and various materials. It is a difficult knot to tie when using two hands, which is why for years we've referred to the half-hitch tool as the fly tier's third hand. This simple tool facilitates quick and easy knot tying and comes in two sizes: one with a tip for #16 and smaller flies, and one with a tip for #14 and larger flies. Many half-hitch tools, like the one pictured here, incorporate the two sizes on the same tool—one on each tip.

Bodkin

The bodkin, or dubbing needle, as it is sometimes called, is simply a needle mounted into a handle. It is used to apply head cement to the head, and sometimes other parts, of a finished fly. It may also be used to "pick out" the fur of dubbed bodies to make a fuzzy-silhouetted fly.

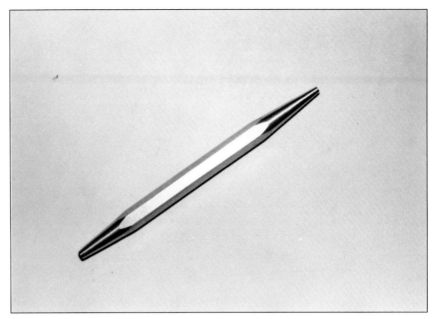

Half-hitch Tool

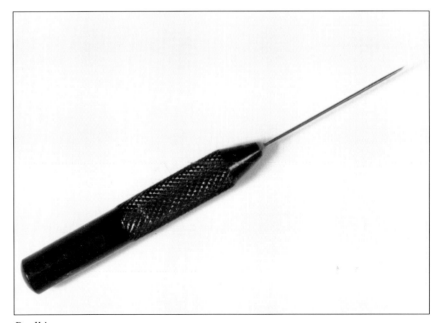

Bodkin

Tools and Materials

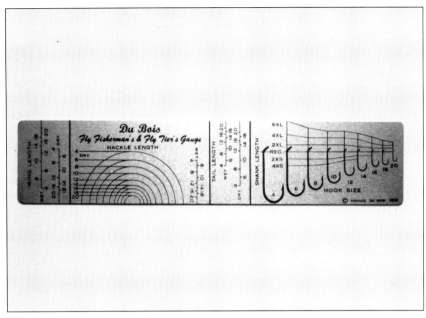
Hackle Gauge

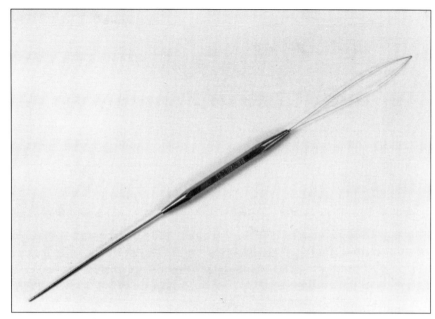
Bobbin Threader

Hackle Gauge

The hackle gauge is a tool used to measure various materials used in tying flies. Some fly-pattern instructions specify precise measurement of materials.

There are at least a half-dozen gauges of varying design on the market. They all work well and are a necessity for the beginning and advanced tier. The gauge in the photograph measures hook, hackle fiber, tail, and wing lengths.

Bobbin Threader

The bobbin threader wasn't around when we started tying some forty years ago, and trying to thread a bobbin then was almost as frustrating as trying to half-hitch or whip-finish (a thread technique to be learned in the next section) with your fingers. The bobbin threader also doubles as a bobbin cleaner: it may be used to clean the bobbin tube of the wax build-up from today's prewaxed threads. There are several styles available and they all work well.

Hooks

Every artificial fly starts with a hook, and there are literally hundreds of styles and manufacturing methods. For the exercises in this book you'll need only three styles of hook, and they will be listed at the end of this section.

There are four major hook manufacturers in the fly-tying world today: Mustad, Tiemco, Belvoirdale, and Partridge. (The Belvoirdale Company is listed in the Appendix and will send a catalog upon request. The other three companies make their hooks available only through retail outlets.) Mustad hooks are appreciably less expensive than hooks from the other three major companies, and for that reason we recommend you use Mustad hooks for the exercises in this book. If you continue in fly tying, however, you may wish to try hooks from the other manufacturers.

Hooks are sized by even numbers in an ascending sequence: the larger the hook, the lower its number. The practical range for flies used in fly fishing is from the large #4 down to the relatively tiny #28. There are hooks larger than #4 and smaller than #28, but they are for special uses and needn't be of concern as you learn to tie.

The photograph illustrates the components of a typical dry-fly hook. Most fly fishers today advocate the use of barbless hooks because they facilitate the release of a fish (not to mention the removal of a hook from an ear or finger). It's a fairly simple operation to turn a barbed hook into a barbless one: simply squeeze the barb with a pair of pliers. Be very careful when doing this on smaller, more delicate hooks, though. It's easy to weaken the metal by squeezing too hard; you don't want to tie flies that are going to break off in the mouth of a trophy lunker.

Hooks are styled for different uses by varying three factors: the thickness of the wire, the length of the shank, and the length of the point and size of the barb. There are other considerations, but these three are the most basic. The look of the fly you tie is directly related to the varying degrees of these three factors in the hook you tie on. The hook's size is a measurement of its gape. The hook's length is the measurement of its effective shank length. Length designations begin with "regular" and become progressively longer through a numbered series from 1XL to 6XL.

This has been an extremely basic explanation of the hook, and we've tried to keep it simple because that's the nature of this book. But as you progress as a fly tier, you will want to learn more about the many kinds of hooks available for specific tying needs.

Here are some commonly used hooks you will need for tying dry flies, wet flies, and nymphs. They are available in packages of 25, 50, and 100.
1. Mustad #94840: Standard wire for dry flies, regular length, sizes 12 to 16.
2. Mustad #3906: Heavy wire, regular length for nymphs, sizes 12 to 16.
3. Mustad #3906B: 1XL for nymphs and wet flies, sizes 12 and 14.

Necks

A "neck" (sometimes called a "cape") is the preserved skin, including the feathers, of the neck and shoulder area of a chicken—a typical barnyard fowl. Necks are divided into two groups: dry-fly necks, which come exclusively from roosters, and wet-fly necks, which come from hens.

Roosters have long, narrow feathers with stiff fibers that are capable of floating a dry fly on the surface of the water. Dry-fly necks are graded according to the stiffness of the hackle fibers as A, B, C, or 1, 2, 3. Grade A or 1 is of the highest quality and will be more expensive than necks of other grades.

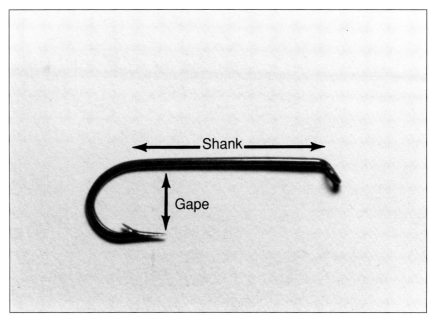

Hooks

The long feathers on the necks of domestic fowl are called *hackle*. There is a small growth of soft fibers close to the stem of each hackle; these fibers pick up and hold water, thus negating the flotation properties of the hackle. Necks with little or no *webbing*, as the soft fibers close to the stem are called, are the most expensive and are graded A or 1. A grade C or 3 neck will have a substantial amount of webbing and will be less expensive.

Wet-fly necks, from hen chickens, have shorter hackle quills with longer and softer fibers and more overall webbing than rooster, or dry fly, hackle. These feathers are ideal for wet flies and some nymphs because they absorb water and lay back around the body of the fly to undulate or "breathe" as the fly drifts in the current. Wet-fly necks are much less expensive than dry-fly necks, whatever the grade.

It's not necessary for a beginning tier to purchase the most expensive, top-grade necks, unless of course he or she wants to. Lesser-grade dry-fly necks will serve adequately while you are learning, and there is one advantage to using them at the beginning of your tying career:

Tools and Materials

you'll recognize and appreciate better-quality hackle once you've used less than optimum material.

The basic colors in wet- and dry-fly necks are black, brown, cream, dun (grayish), and grizzly (dark and light barred). There are many other colors (some dyed), patterns, and variations, but these five will serve you well as you begin.

Thread

Today's fly-tying thread is vastly superior to the threads of past years. It is strong, easy to work with, and in most cases comes from the manufacturer with a light coating of wax to make winding easier and adhere better to various materials. For dry-fly work, regular thread in size 6/0 is used. There are threads finer and thicker than the size 6/0, but it is the standard in fly tying. For heavier work on wet flies and nymphs, either nymph thread or a product called Monocord is used. You will notice the difference in tensile strength between the 6/0 and nymph thread very quickly; learning how much tightness a particular thread will take without snapping is one of the skills the beginning tier needs to pay particular attention to.

The basic colors in black, brown, white, and olive in both regular 6/0 tying thread and nymph thread or Monocord will see you nicely through the beginning stages of learning to tie.

Head Cement

Head cement is a clear lacquer used on fly heads to make them hard and durable. (It is also used for coating wings, wing cases, and underbodies.) There are many brands available and every tier seems to have his own particular favorite. Be sure you have adequate ventilation when using head cement—it can be potent stuff!

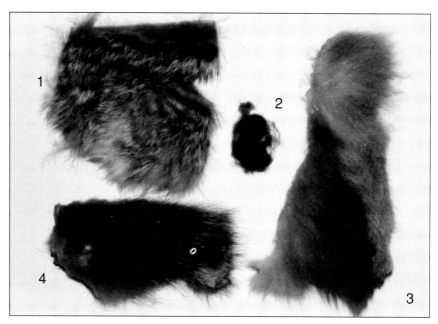

Some common dubbing materials: 1) rabbit fur, 2) poly yarn, 3) red fox, 4) opossum.

Dubbing Fur

"Dubbing" is a process whereby bits of fur are made to adhere to prewaxed tying thread. The dubbed thread is then wound around the hook to fashion a somewhat fuzzy fly. There are many furs and even synthetic materials used as dubbing. You can buy these materials preblended or blend your own to obtain desired shades of color. Preblended dubbing fur looks much like a big, rough, cotton ball. For the exercises in this book you will need dubbing material, either natural fur or synthetic. Among the most common furs used for dubbing are muskrat, rabbit, red fox, and opossum.

Different types of hair: 1) squirrel tail, 2) caribou, 3) bucktail, 4) deer body hair.

Various feathers: 1) turkey tail, 2) pheasant tail, 3) turkey wing, 4) duck wing, 5) peacock eye, 6) marabou, 7) wood duck, 8) neck.

Hair

Hair from various animals, notably the deer, is widely used in fly tying. For the exercises in this book you will use deer body hair in brown or natural and dyed black, and natural bucktail. Deer body and facial hair differs from the hair found on the tail of deer (called "bucktail"). In general, the finer-textured hair from the bucktail is used for wings, and the thicker hair from the deer's body is used for trimmed-hair bodies and tails. Deer hair is extremely buoyant because it is hollow and is especially useful in the construction of dry flies used in heavy water conditions.

Feathers

All kinds of feathers are used in various ways to create artificial flies; feathers are really the foundation of the craft of fly tying. Various feathers have widely diverging shapes and textures, as you will learn as you continue to tie. For the exercises in this book you will use, in addition to the chicken hackle described earlier: mallard wing feathers, wood-duck feathers, and peacock herl (herl is the fuzz found on certain feathers, in this case, the fronds of a peacock's tail).

Synthetic Materials

In the last twenty years or so, synthetic materials for fly tying have appeared at an amazing rate. The synthetics you will need for the exercises in this book include foam cylinders, which we find just wonderful for shaping the bodies of terrestrial insect imitations: ants and beetles, for instance. You will also use chenille material to create a wrapped body, tying tinsel to create ribbing, and poly yarn for wings.

Tools and Materials

Basic Techniques

Part 2

- Threading the Bobbin
- Placing the Hook in the Vise
- Tying the Half-Hitch
- Tying the Whip-Finish
- Weighting a Hook
- Alternative Weighting Method
- Applying Head Cement

The exercises in this section will teach you basic operations that are performed, sometimes more than once, each time you tie a fly. Almost all of them have to do with managing the thread and bobbin and we recommend you practice each one until you feel comfortable and adept. The extra time you spend now, becoming comfortable with the feel of the thread and learning its limits, will give you confidence as you begin to add different materials to fashion the various components of a fly.

One thing we'd like to recommend you do as you go through the steps in the exercise "Tying On Thread" is to break the thread. Yes, you read that correctly—break the thread. All fly tiers do this occasionally, even the most experienced. We think beginning tiers should set out to do so deliberately, because nothing is better at teaching how much tension thread will take than applying too much right off the bat. Once you've learned how much is *too* much, you're well on the way to recognizing the "feel" of how much is just right.

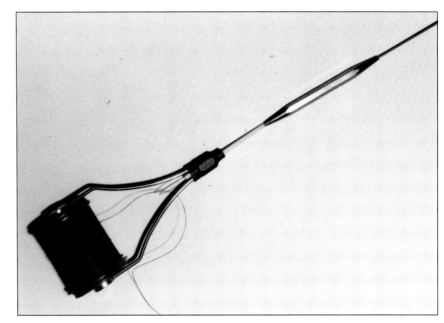

Step 1, Threading the bobbin

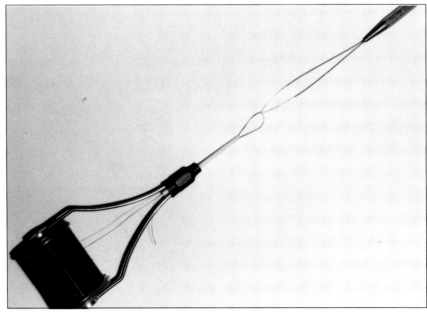

Step 2, Threading the bobbin

Threading the Bobbin

Threading a bobbin is simple when you use the bobbin threader.

Step 1: Secure the spool of thread in the bobbin by pulling the bobbin arms apart and clipping one on each end of the spool. Adjust the bobbin to hold the spool firmly. Insert the wire tip of the bobbin threader into the bobbin's tube, toward the spool. When the wire is through the tube, pass the end of the thread through the wire as if you were threading a needle.

Step 2: Now pull the wire attached to the thread back out of the bobbin tube. Pull the thread out of the threader and your bobbin is ready to use. There is no tension in the bobbin tube to hold the thread—pull enough thread through the stem so that it can't easily slip back.

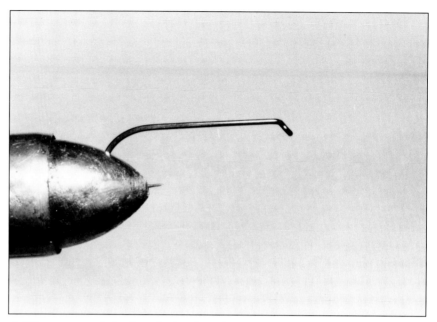
1. Wrong way of placing the hook in the vise.

Placing the Hook in the Vise

1. This is how *not* to place a hook in the vise. The top of the vise jaws will put a kink in the bend of the hook, thus weakening the metal. Such a hook will be likely to break under the stress of being caught on twigs, fish, or rocks.

2. This is the correct way to secure a hook in the vise. The vise-jaw pressure is distributed evenly across the bend of the hook. Be careful not to tighten the jaws around the hook too much. Even if the hook is placed correctly, too much jaw pressure will weaken the bend.

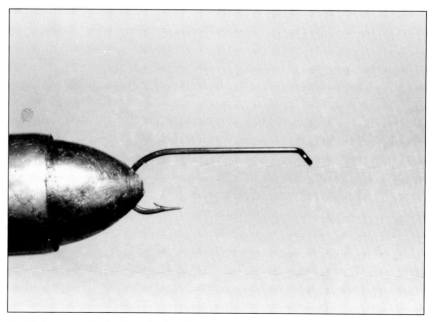
2. Right way of placing the hook in the vise.

Basic Techniques

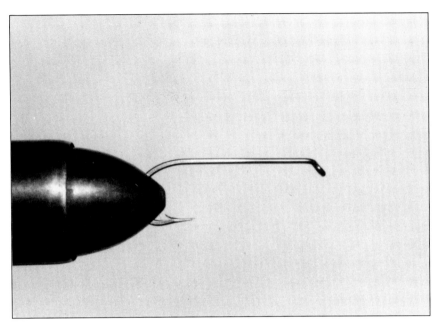

Step 1

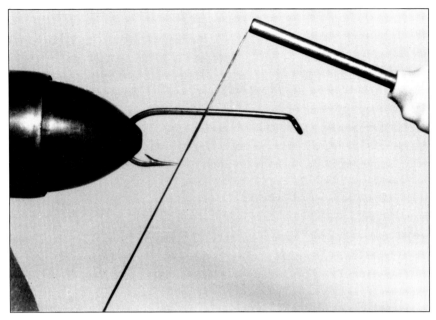

Step 2

Tying On Thread and Tying the Half-Hitch

Tying the thread on the hook is the first step in any fly pattern. The half-hitch is a knot that is used over and over again in fly tying. It is used after various tying operations are completed to secure the thread or materials.

We recommend you practice this exercise on a relatively large hook until you're comfortable. Try a #12 or #14 regular length or 1XL dry-fly hook. Use nymph thread or Monocord at first—they're thick and easier to tie with. Then, when you're familiar with the technique, try it with regular 6/0 tying thread, which is thinner than nymph thread, and a little harder to handle.

Step 1: Insert the hook in the vise properly.

Step 2: Hold the loose end of the tying thread below the hook in your *left* hand. With your *right* hand, pull the bobbin stem (and the thread) above the hook. Lay the line of thread against the side of the hook that faces you, as shown in the photograph. The thread should be slanted slightly, and the lower end should pass just in front of the point of the hook.

HALF-HITCH

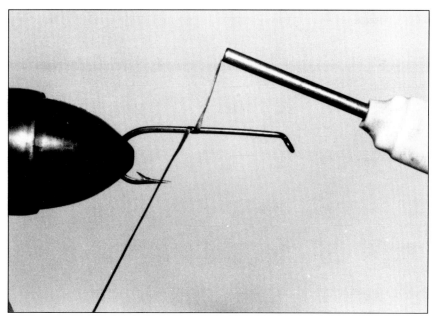
Step 3

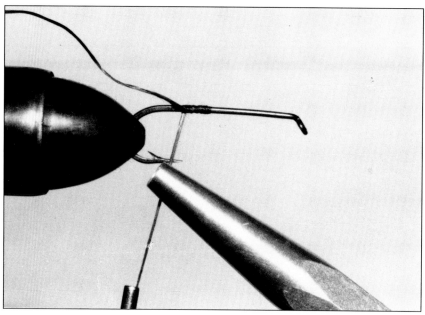
Step 4

Step 3: Move the bobbin away from you over the top of the hook. Keep moving the bobbin over the hook top, then down the other side of the hook, then underneath the hook and back toward you, completing a circular motion. When the bobbin stem is back in the starting position, as pictured, you've made one forward wrap, or turn, of thread. Continue to hold the thread end in your left hand.

Step 4: Make six to eight more wraps, but this time wrap *backward*, in other words toward the rear of the hook. To do this you will wrap over the first wrap and over the thread you are holding in your left hand. As you make these backward wraps, pull the thread in your left hand toward the vise, as shown in the photograph, to keep it out of the way. Each successive wrap should abut the previous wrap snugly. You should be applying enough tension with the bobbin to keep your wraps smooth and tight, but not so much that you break the thread! Practice will teach you how much tension to apply. As your wraps come closer to the hook point, you will have to wrap at a slight angle to avoid fraying or breaking the thread on the sharp point. Maintain a constant tension as you angle your wraps around the hook point. After you've completed the six to eight wraps, apply a half-hitch to secure them. To begin the half-hitch, let go of the loose end of the thread (it's not going anywhere) and switch the bobbin to your left hand. Take the half-hitch tool in your right hand. Pull the thread down and slightly to the left, as pictured. Lay the upper part of the half-hitch tool against the thread, keeping the tool between the thread and you.

Basic Techniques

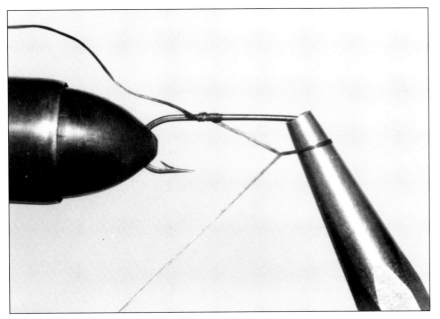

Step 5

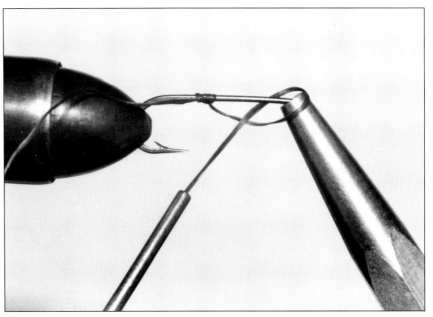

Step 6

Step 5: Make one wrap of thread around the half-hitch tool by bringing the bobbin toward you and moving the thread around the tool. Place the tip of the half-hitch tool over the eye of the hook, applying enough tension to keep the thread taut.

Step 6: With the tip of the half-hitch tool in position over the eye of the hook, slide the loop of thread toward the top of the half-hitch tool by firmly and evenly pulling the bobbin up slightly and to the left. Continue to pull the loop of thread over the top of the half-hitch tool and over the hook eye onto the shank.

HALF-HITCH

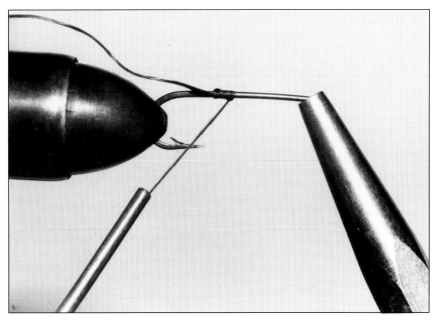
Step 7

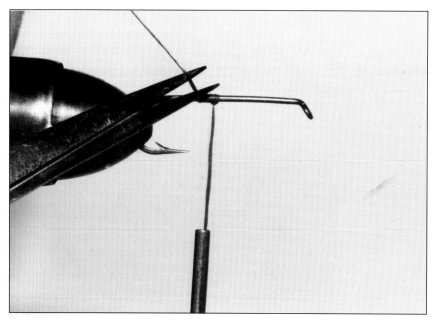
Step 8

Step 7: Slide the loop along the shank until it reaches the end of the thread wraps. Tighten the loop firmly against the thread wraps. Your single half-hitch is complete.

Step 8: Pull the dangling thread end upward in your left hand. Clip off this excess thread as close to the hook as you can, being careful not to cut into the wraps.

Step 9: Your thread is now tied on and you are ready to begin constructing a fly according to whatever pattern you've chosen. The tie-in position is almost always toward the rear of the hook shank, as shown in the photograph.

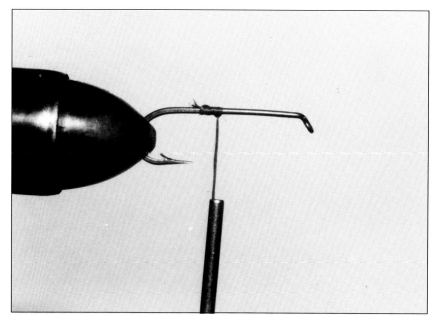
Step 9

Basic Techniques

Tying the Whip-Finish

The whip-finish is a knot used to tie off the head of a finished fly. There are several ways to execute a whip-finish, but the method shown here is by far the easiest. Practice until you feel comfortable tying it.

To practice the whip-finish, use the hook and thread you did in the previous exercises for tying on and tying the half-hitch.

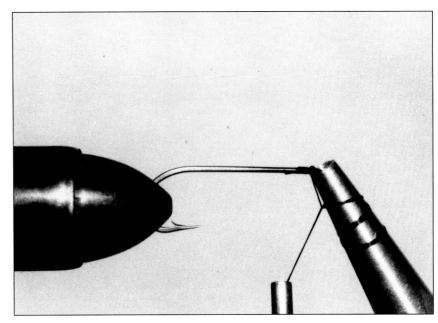

Step 1

Step 1: Secure the hook in the vise and tie on near the eye of the hook; when you're whip-finishing a fly you've completed tying, this is where the thread will be positioned. Just as in tying the half-hitch, hold the bobbin in your left hand so the thread is angled down and to the left of the hook's eye, and with your right hand place the half-hitch tool against the tying thread, keeping the tool between you and the thread. Make *three* turns of thread around the tip of the tool. Applying enough tension to keep the thread taut, slip the end of the half-hitch tool over the eye of the hook.

Basic Fly Tying

WHIP-FINISH

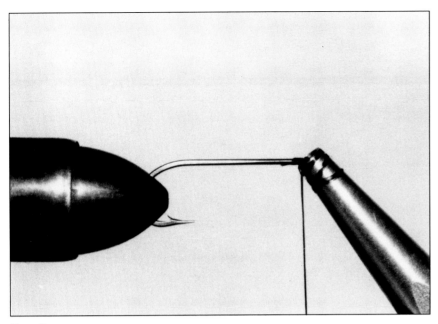

Step 2

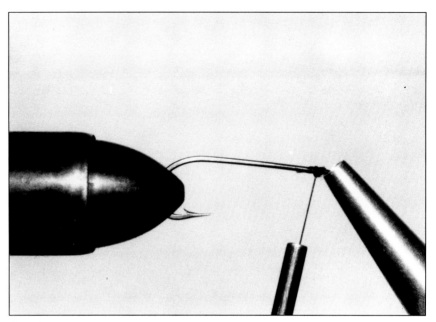

Step 3

Step 2: With the tip of the half-hitch tool firmly in position over the eye of the hook, pull the thread you are holding in your left hand firmly and evenly off the half-hitch tool and onto the hook shank.

Step 3: Pull the thread tight against the thread wraps (or head of a finished fly). Regular tying thread is fine enough that you should not create a big lump from a whip-finish knot. Repeat the process two more times to complete a whip-finish on the head of the fly.

Basic Techniques

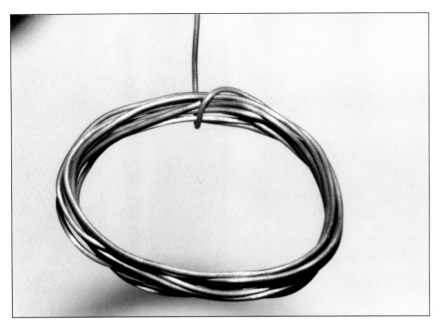
Lead Wire

Step 1

Weighting a Hook

Sometimes a pattern calls for adding weight to a fly to increase its effectiveness in fast, heavy, or deep water. It may be a wet fly, nymph, or streamer. Adding weight to a hook is simple. All you need is a coil or spool of soft lead wire in $1/64$-inch or $1/32$-inch diameter. (In the photographs, $1/32$-inch wire is used.) Lead wire can be purchased from fly-tying shops, mail-order catalogs, or hardware stores. Weighted flies are often tied using a different color thread from that used on a fly with an unweighted hook. This enables you to distinguish between your weighted and unweighted flies.

Step 1: Cut a 4- or 5-inch length of lead wire from the coil or spool. (The length you use will depend on the size of the hook and how heavily you want the fly weighted.) Insert the hook in the vise. With the thumb and forefinger of your right hand, hold the end of the wire against the forward part of the hook shank (toward the eye). With your left hand make two or three turns toward the rear of the hook. You'll find the lead bends very easily.

WEIGHTING A HOOK

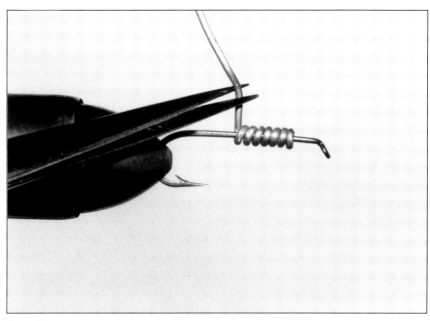

Step 2

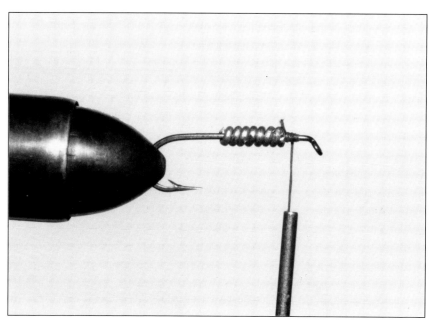

Step 3

Step 2: Hold the wraps you've just make with your left hand to keep them in place. With your right hand, wrap the short length of wire forward on the hook; it may take four to six turns to use up all the lead wire. Clip off the excess lead at the rear of the shank. Wrap and pinch it down to make a neat, even foundation on which to tie your weighted fly.

Step 3: Tie on with six or eight turns of thread just in front of the lead wire and behind the eye of the hook. Pull the thread taut, apply a half-hitch, and clip off excess thread.

Step 4: Wrap the thread backward over the turns of lead wire and down onto the hook shank behind the last wrap of lead. Apply a half-hitch. You are now ready to begin tying a fly pattern requiring a weighted body.

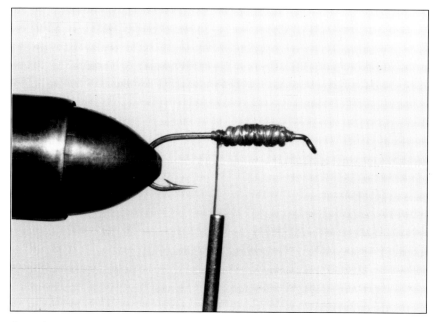

Step 4

Basic Techniques

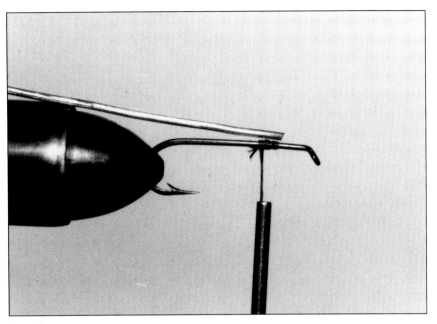

Step 1

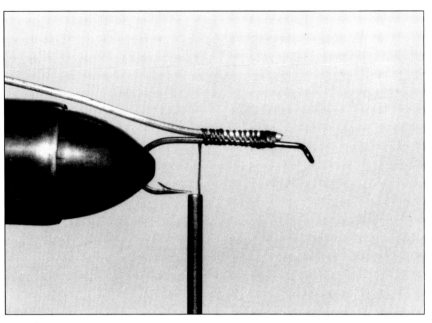

Step 2

Alternative Weighting Method

A single strip of lead wire may be tied on top of the hook shank when less weight is required. This method produces a slenderer body.

Step 1: Tie on near the front of the hook. Hold the lead wire in your left hand and place it on top of the hook shank.

Step 2: Hold the wire firmly in place with your left hand. With the bobbin in your right hand, wrap the thread tightly around the wire and hook shank toward the rear of the hook.

Step 3

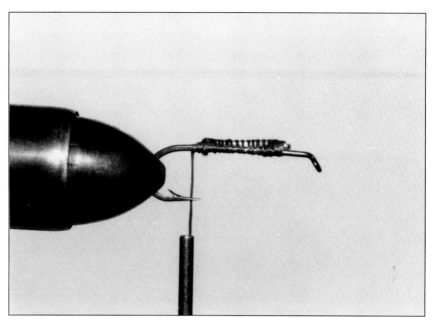

Step 4

Step 3: Clip off the wire. The length of wire you use will depend on the amount of weight needed for the specific fly pattern you want to tie.

Step 4: Continue to wrap the thread over the clipped end of the lead wire and down onto the hook shank. You are now ready to tie your weighted nymph, wet fly, or streamer.

Applying Head Cement

Head cement is a clear lacquer applied to the head of a finished fly to help strengthen and preserve it. It is very easy to apply. Simply use your bodkin or the applicator that comes with the bottle of head cement and place one drop directly onto the fly's finished head. Allow the head cement a few minutes to dry completely, then clear any excess cement out of the hook eye.

Tails

Part 3

- Hackle-Fiber Tails
- Deer-Hair Tails
- Split Tails

Tails are an important component of a well-balanced fly. In addition to imitating properly the silhouette of a natural insect, tails on a dry fly serve two important functions: they help the hook light on the water in correct position (point down) after being cast, and they hold the aft-end of the hook up in the water to aid in proper flotation.

Various materials are used for the tails on nymphs, wet flies, and dry flies. Two of the most commonly used materials for tails—hackle fibers and deer hair—are used in the exercises for the three tails covered in this section. Two of the tails are straight and one is split into two tail sections. When you've mastered these techniques you will be able to tail any fly pattern correctly and with ease.

Step 2

Hackle-Fiber Tail

For this exercise you will need a hackle from a wet- or dry-fly neck. Any of the five basic colors mentioned in Part One will be satisfactory (we are using grizzly hackle in the photographs). We suggest you use a #12 or #14 regular length hook and regular 6/0 tying thread. (When you tie from a pattern you will follow its instructions for materials to be used.)

HACKLE-FIBER TAIL

Step 3

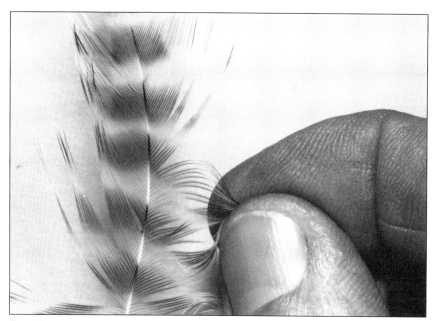
Step 4

Step 1: Secure the hook in the vise and tie on. (Refer back to page 20 and the exercise "Tying On Thread" if you need a refresher.)

Step 2: From the neck, select one of the larger hackles from the middle or upper section of the neck. (Larger hackle is used for tails.) Between the thumb and forefinger of your right hand, pull the hackle you've selected out to the side of the neck, as pictured.

Step 3: Applying pressure near the base of the hackle with your thumb and forefinger, pull the hackle stem from the neck.

Step 4: Hold the tip of the hackle stem in your left hand and firmly pinch a bunch of the hackle fibers from one side of the stem between the thumb and forefinger of your right hand. The correct amount for this size hook is approximately one-half to three-quarters of an inch in width, as pictured.

Step 5: Now pluck the bunch of fibers off the stem with an easy stripping motion. (They should come off with little resistance.)

Step 5

Tails

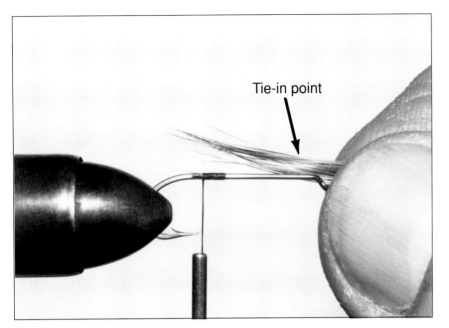

Step 6

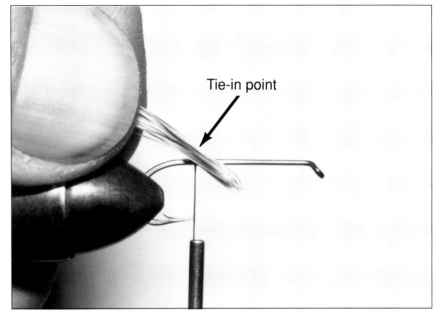

Step 7

Step 6: Holding the butts of the hackle fibers between the thumb and forefinger of your right hand, measure the fibers against the straight portion of the hook shank for proper length. The tail when finished and trimmed should be as long as the straight part of the shank. Make a mental note of the correct tail length, when measured from the tips toward the butts, before you tie the fibers down. The point you have noted (it should be closer to the butts than the tips) is the point on the fibers where you will tie them onto the hook. (You can use a ruler for exact measurements if you wish, but it's not necessary to be *that* precise. Almost all fly tiers eyeball tail length.)

Step 7: Transfer the fibers from your right to your left hand by grasping the tips with the thumb and forefinger of your left hand. Hold the hackle fibers at the slight angle shown in the photograph against the *side* of the hook that faces you. The point on the fibers you have noted as the correct tail length should be the point that rests against the hook.

HACKLE-FIBER TAIL

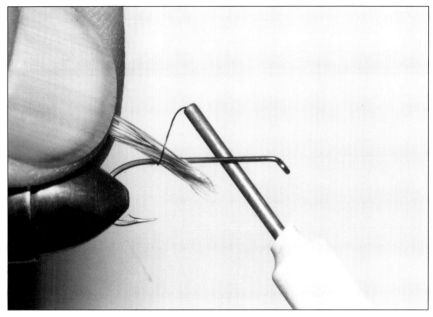

Step 8

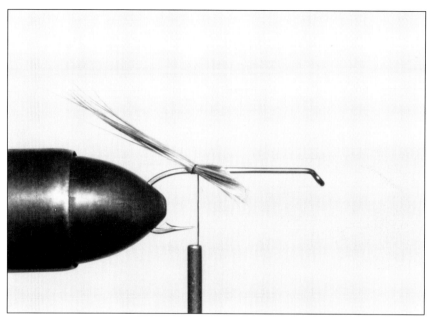

Step 9

Step 8: With the hackle fibers correctly positioned on the side of the hook in your left hand, pick up the bobbin in your right hand. Bring the bobbin, which should be hanging down from the off-side of the hook, under the hook up toward you and over the top of the hook to the off-side, completing one *loose* thread wrap, as pictured.

Step 9: Pull tight the loose wrap you've just made and continue to hold the hackle fibers with your left hand. The fiber bunch should begin to shift toward the top of the hook shank, as pictured.

Tails

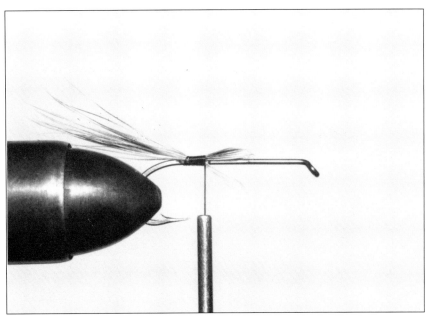

Step 10

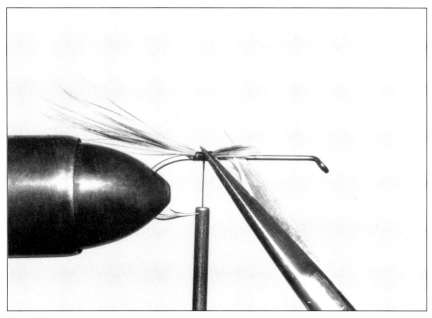

Step 11

Step 10: Make four to six wraps of thread forward, over the hackle fibers, toward the eye of the hook. Be sure to pull each wrap snug and abut it smoothly against the previous wrap. With each wrap the tail fibers should move closer to their final position *on top* of the hook shank. When the fibers sit on top of the shank, as pictured, you've wrapped enough. (Don't rely on a specific number of wraps, which are given only as guidelines. When the fibers reach their final position on top of the shank you can stop wrapping, no matter how many turns of thread it takes.) Apply a half-hitch, and when you pull it over the eye of the hook onto the shank, slide it along the shank *underneath* the fiber butts and up against the thread wraps before you pull it tight. (Refer back to page 21 and the exercise "Tying the Half-Hitch" if you need a refresher.) Let the bobbin hang from the off-side of the shank.

Step 11: Clip the butts of the hackle fibers close to the thread wraps, but be careful not to cut the thread.

HACKLE-FIBER TAIL

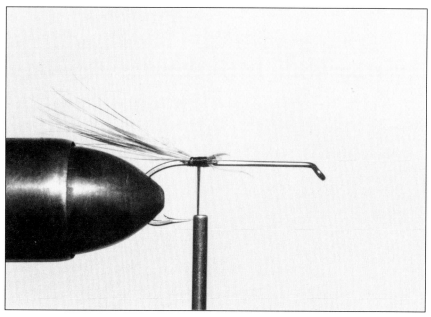

Step 12

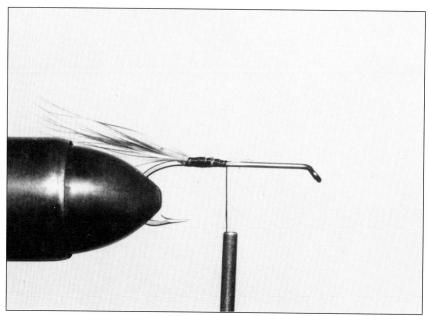

Step 13

Step 12: The trimmed ends don't have to be perfectly neat. As you can see from the photograph, we've left some ragged ends. If you were following a specific pattern, the body of the fly you were constructing would cover the cut ends of the tail fibers.

Step 13: Pick up the bobbin in your right hand and wrap forward just enough to cover the fiber ends.

Step 14: Apply a half-hitch and pull it snug against the edge of the thread wraps you've just made. Your hackle-fiber tail is now complete. (If you were following a fly pattern, at this point you would continue on to construct the fly body.)

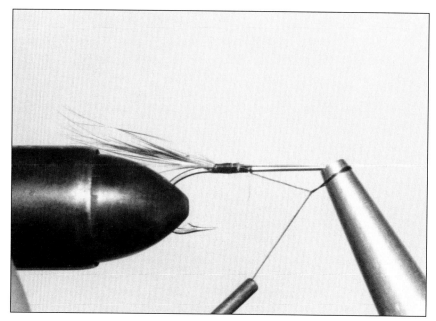

Step 14

Tails

Deer-Hair Tail

For this exercise you will need a patch of deer body hair in black or brown and a #12 or #14 regular dry-fly hook. The thread type and color you use normally depends on the materials used for the various components of the fly. For this exercise, use nymph thread as a compromise between the weaker 6/0 thread and the heavy Monocord.

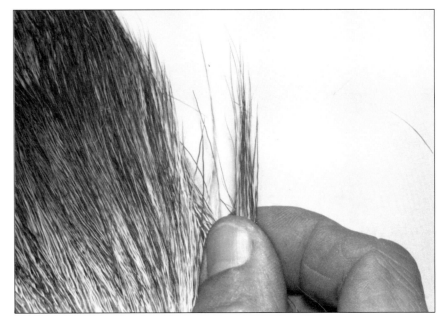

Step 2

Step 1: Secure the hook in the vise and tie on.

Step 2: From the patch of deer body hair, select a clump that is roughly the thickness of a soda straw.

DEER-HAIR TAIL

Step 3

Step 4

Step 3: Take the tips of the clump of deer hair between the thumb and forefinger of your left hand. Pull the clump straight out from the hide so that the tips are even.

Step 4: With your largest pair of tying scissors, cut the clump of hair close to the hide.

Step 5: Hold the clump of deer hair by the cut ends between the thumb and forefinger of your right hand and measure it against the straight part of the hook shank for proper length. The finished tail should be as long as the straight portion of the shank. (As you learned in tying the hackle-fiber tail, make a mental note of the correct length before you tie the deer-hair fibers onto the hook. The point you note is the point on the fibers to tie to the shank.)

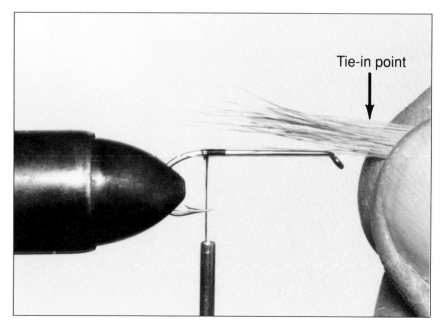
Step 5

Tails

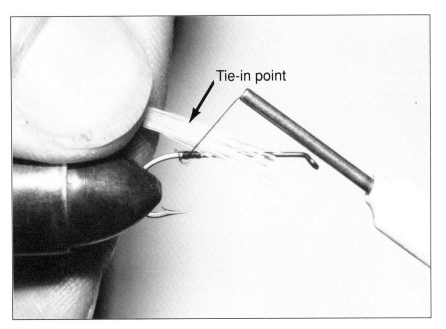

Step 6

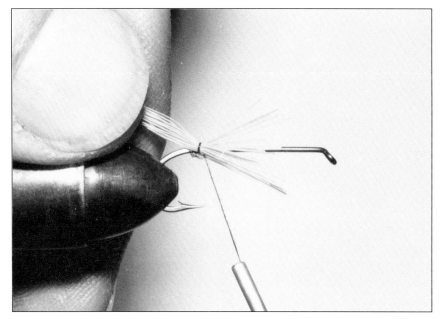

Step 7

Step 6: With your left thumb and forefinger, grasp the hair fibers by the tips. Position the fibers at the slight angle shown in the photograph against the side of the hook that faces you. The point on the fibers that you have noted as the correct tail length is where the fibers should rest against the hook. Pick up the bobbin with your right hand and make one loose thread wrap. (As you no doubt recognize, the procedure so far is the same as for the hackle-fiber tail.)

Step 7: Pull the thread wrap tight. It is important that you continue to hold the fibers with your left hand so the hair does not slip. Because deer hair is stiff and somewhat slippery, the tail will be in danger of slipping until all the thread wraps are completed. Learning to manipulate deer hair successfully is one of the trickier skills in fly tying. Try not to become discouraged if your deer-hair tails slip the first few times you try to tie them. Simply unwind the thread, remove the hair, and begin again.

Step 8: As you continue to hold the hair fibers with your left hand, make six to eight thread wraps forward, until the tail shifts completely to the top of the hook shank. (As you will remember from the hackle-fiber tail, each wrap moves the tail closer to the top of the shank.) When the tail is on the top of the shank and wrapped securely, you can let go of the fibers with your left hand.

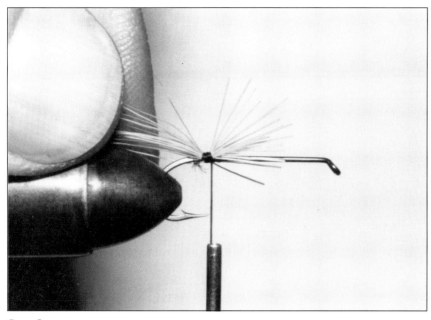
Step 8

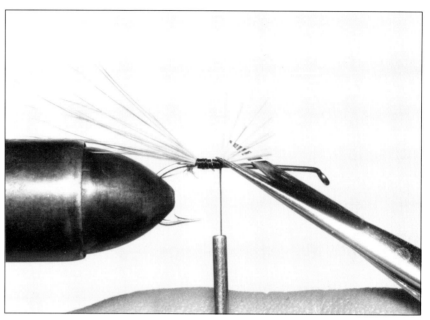
Step 9

Step 9: Carefully spread the butts of the tail fibers up and back away from the hook shank (this makes them easier to trim). Apply a half-hitch, sliding it over the eye of the hook, down the shank and under the butts of the tail fibers, and snug up against the thread wraps. Let the bobbin hang from the off-side of the hook shank. With your large scissors, clip off the excess butt ends, trimming carefully close to the thread wraps. Try to cut off the butt fibers on an angle; this will result in a smoother base for the body.

Step 10: Wrap forward just enough to cover the cut ends of the hair fibers. Apply a half-hitch. Your deer-hair tail is now complete. (As with any tail, you would continue on at this point to construct the rest of the fly according to the pattern you were following.)

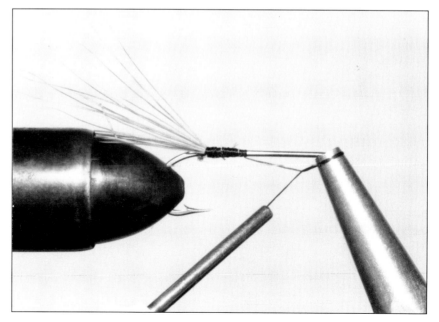
Step 10

Split Tail

This sequence is photographed looking *down* on the fly in order to illustrate the tying techniques effectively. We're also holding the thread and bobbin out to the side so you will have a better view of its position; you would not do this when tying this tail.

For this exercise you will need a #12 or #14 regular length dry-fly hook, regular 6/0 tying thread or nymph thread, and a bucktail. (This method also works well with rooster hackle fibers or guard hairs from mink tail, muskrat, or beaver in place of the bucktail.)

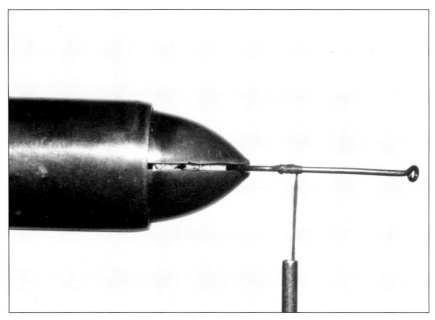

Step 1

Step 1: Secure the hook in the vise and tie on in the regular starting position toward the hook bend. Let the bobbin hang.

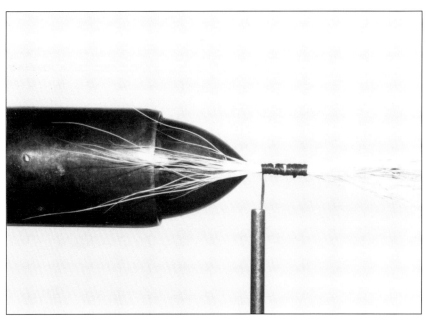

Step 2

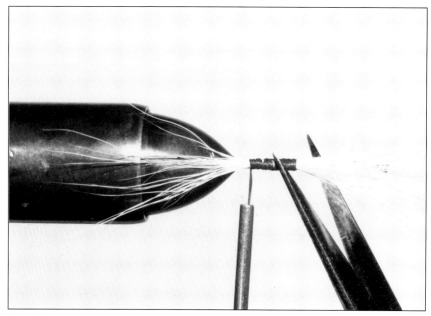

Step 3

Step 2: As you did in the previous exercise, "Deer-Hair Tail," select a clump of hair from a bucktail of roughly the thickness of a soda straw and measure it for correct length against the straight portion of the hook shank (the finished tail should be about the length of the straight portion of the shank). Holding the hair fibers on *top* of the shank, make six to eight thread wraps forward around the hair to secure it and apply a half-hitch. For this tail, the fibers should begin and remain on *top* of the hook shank.

Step 3: When the fibers are tied in securely, let the bobbin hang and trim the excess fiber ends close to the wraps, being careful not to cut into the thread.

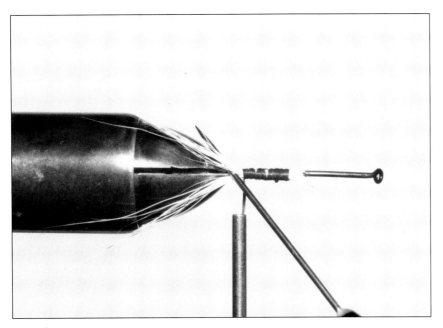

Step 4

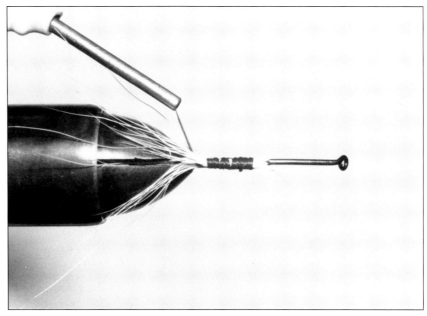

Step 5

Step 4: Use your bodkin to split the tail fibers in half, as pictured, pushing each half out to the side at the slight angle shown. The tail fibers will be held in this position using the figure-eight method described in the following steps.

Step 5: Pick up the bobbin in your *left hand* and bring it up from underneath the hook shank and toward you. Keep moving the bobbin over the shank and bring the thread down between the split fibers of the tail and to the off-side of the hook shank. Put down the bodkin and switch the bobbin to your right hand. Make three or four more of these diagonal wraps, bringing the thread up toward you, then over and down between the split tails to the hook's off-side. On the last wrap, don't bring the thread through the split; just make a circular wrap around the hook. This secures the diagonal wraps you've just done. At this point you have completed one-half of the figure eight.

SPLIT TAIL

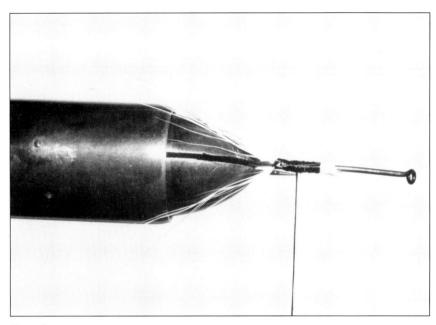

Step 6

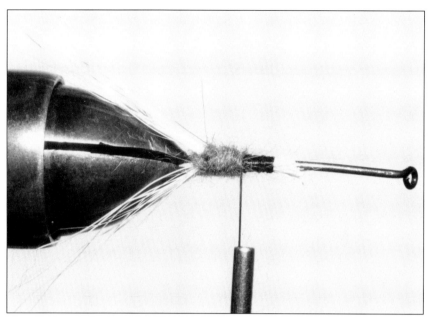

Step 7

Step 6: Bring the thread under the hook shank and up between the hook shank and the tail fibers nearest you. Take it through the split, then down *in front* of the tail fibers on the off-side of the hook. You have just made a diagonal wrap in the opposite direction. Make three or four more exactly like that. This completes the figure eight.

Step 7: Now you will spin fur on the tying thread, following the instructions in the "Dubbed Fur Body" exercise on page 48. Repeat steps 5 and 6, except make only one diagonal wrap in each step. The thickness of the dubbed thread helps keep the tail sections apart. This technique is used with all types of fly bodies requiring split tails, not just with dubbed bodies.

Tails

Bodies

Part 4

- Dubbed Fur Body
- Dubbed Fur Body with Tinsel Rib
- Chenille Body
- Peacock Herl Body
- Clipped Deer-Hair Body
- Quill Body
- Thorax and Wing Case
- Foam Body

The body of an artificial fly is the component that most influences the finished shape. Many types of natural and synthetic materials are used to create different types of fly bodies. The dubbed fur body is perhaps the most common on dry flies. The dubbing material is "spun" directly onto the tying thread and then wrapped on the hook shank to create a somewhat bulky or fuzzy silhouette. Winding chenille or peacock herl onto the hook is another way to create a fuzzy body and is popular in nymph imitations.

Bodies constructed with the stem (called "quill") of various feathers are good representations of the segmented bodies of many kinds of insects. Quill bodies differ mainly in the types of feathers used and in the manner the fibers are stripped from the feathers to expose the quill. In this section you will tie a quill body using a hackle stem from a wet-fly neck. Tinsel is also used to "rib" a body and you will learn how to wind tinsel for ribbing in this section.

Deer hair, because it is hollow, has excellent floating properties. In some patterns, deer-hair fibers are tied onto the hook parallel to the shank. In this section you will learn how to "flare" deer hair onto a hook and then clip it at right angles to the shank to create a deer-hair beetle.

A number of fly patterns require that a "thorax," corresponding to the area in back of the head of some insects, be tied. In many nymph imitations the thorax is a base for adding wing cases. In this section you will use a quill body as a starting point for constructing a dubbed-fur thorax, used in some nymph patterns.

For years fly tiers created imitation ants by patiently tying dubbed fur or poly-yarn bodies. The relatively recent introduction of closed-cell foam material from which the ant body can be constructed offers an easy, effective alternative. You will tie a foam-bodied ant in this section, preparing you to go on to tie other terrestrials with the same material.

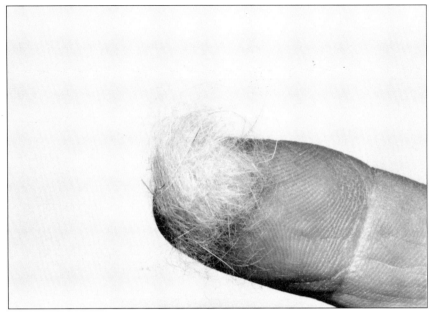

Step 2

Dubbed Fur Body

For this exercise you will need to tie a straight tail, either hackle-fiber or deer-hair, on a #12 or #14 regular length hook. Use 6/0 tying thread. In addition, you will need dubbing fur.

DUBBED FUR BODY

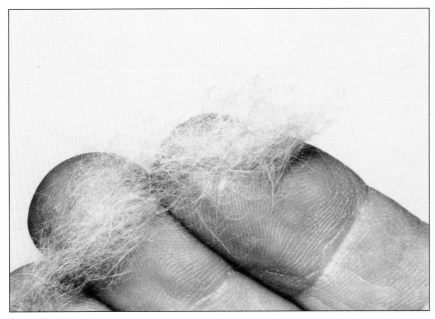

Step 3

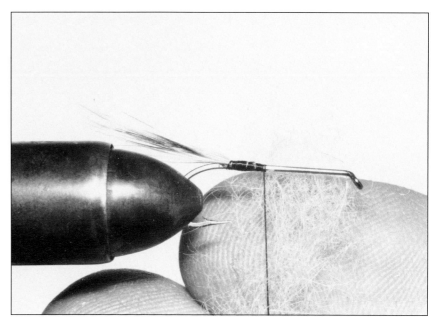

Step 4

Step 1: Secure the hook in the vise and tie on a straight tail, following the instructions for either a hackle-fiber tail (page 32) or a deer-hair tail (page 38). Let the bobbin hang down the off-side of the hook.

Step 2: Select a small clump of dubbing fur about the diameter of a dime.

Step 3: Gently tease the fur apart with your fingers to a length of about two inches, as pictured. Lay this length of fur on the first three fingers of your right hand.

Step 4: Lay the fur with the fingers of your right hand against the off-side of the hanging thread, just below the hook. (If your thread is not pre-waxed, wax it before putting the fur in place.)

Step 5: Use the thumb of your right hand to gently pinch the fur and thread against your fingers. As in the picture, the dubbing fur should stick to the thread.

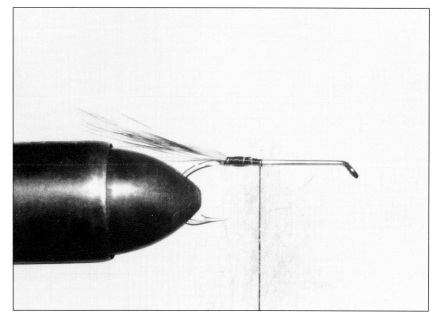

Step 5

Bodies

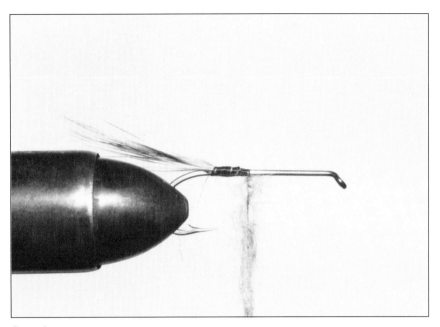

Step 6

Step 7

Step 6: With the thumb and first two fingers of your right hand pinching the fur against the thread, gently roll *in one direction*, working down the thread from the hook shank until you've spun the entire length of dubbing fur onto the thread. When you're finished, the dubbing fur should be spun tightly onto the thread, as pictured. The thread is now "dubbed."

Step 7: Pick up the bobbin in your right hand and begin wrapping the dubbed thread backward, toward the rear of the hook shank, to the last wrap of thread that ties in the tail.

Step 8: As soon as you've made your last backward wrap, begin to wrap the dubbed thread forward over the backward wraps you've just taken. The body you're creating should be chunky because of the dubbed thread.

Step 9: As you wrap forward, if you reach the end of the dubbed thread, simply stop wrapping, allow the bobbin to hang, and spin more dubbing onto the thread. With practice you will be able to judge the correct amount of dubbing the first time.

Step 10: Continue wrapping the dubbed thread forward on the hook shank to a point about a quarter of an inch behind the eye of the hook. This leaves room for the rest of the fly.

Step 11: If you reach the stopping point and still have fur dubbed onto the thread that is hanging beneath the shank, pull the dubbing off so that you can apply a half-hitch with undubbed thread. Apply the half-hitch, pulling it snug against the end of the dubbed body, which is now complete. (If you were following a fly pattern, you would go on at this point to add other components to the dubbed body.)

DUBBED FUR BODY

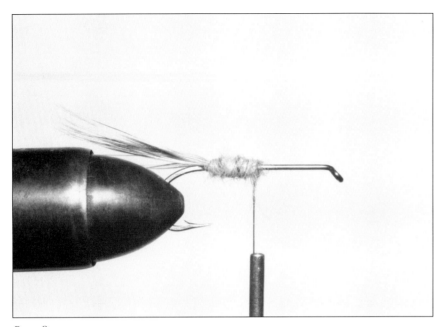

Step 8

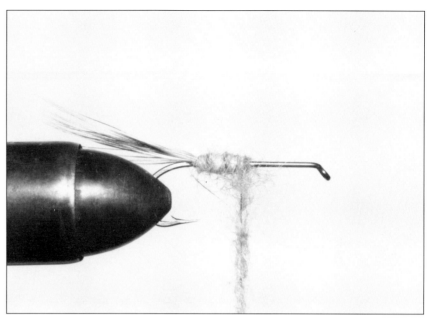

Step 9

Step 10

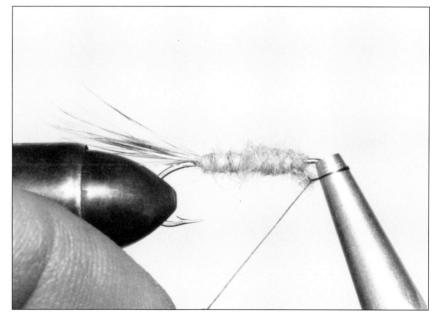

Step 11

Bodies

51

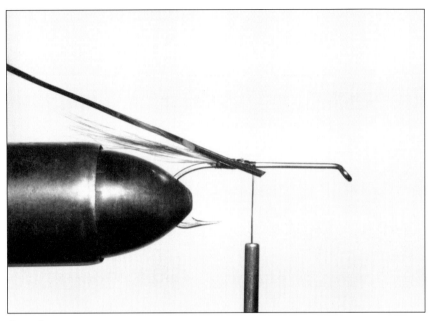

Step 3

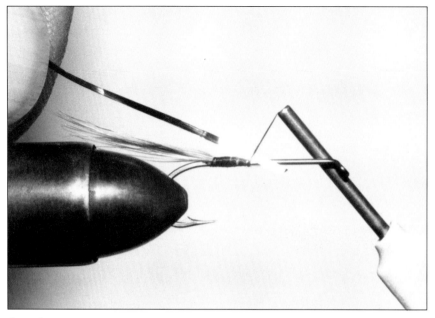

Step 4

Dubbed Fur Body with Tinsel Rib

For this exercise you will construct a straight tail of your choice (hackle-fiber or deer-hair) on a #12 or #14 regular length hook using 6/0 tying thread. In addition, you will need dubbing fur and fly-tying tinsel.

Step 1: Secure the hook in the vise and tie on a straight tail.

Step 2: Cut a four-inch length of tying tinsel.

Step 3: Hold one end of the tinsel in your left hand and position the other end at a slight angle against the side of the hook that faces you, as pictured. (This is the same principle as for tying in tail materials.)

Step 4: Continue to hold the tinsel in your left hand, and with your right hand begin one loose turn of thread around the tinsel end and the hook shank, leaving a tag-end of tinsel that is about one-half inch or so in length.

DUBBED FUR BODY WITH TINSEL RIB

Step 5

Step 6

Step 5: Complete the wrap and pull it tight. The tinsel will move slightly toward the top of the hook.

Step 6: Continue wrapping (about six to eight more turns) forward, until the tinsel is in position on top of the hook shank. When the tinsel is tied in securely you can let go of the other end with your left hand. Let the bobbin hang down the off-side of the shank and clip off the tinsel tag-end close to the thread wraps, being careful not to cut the thread.

Step 7: Pick up the bobbin and wrap six to eight turns of thread backward, toward the rear of the hook, over the previous forward wraps. (We're holding the tinsel end up so you can see what's happening in the photographs. It's not necessary for you to do this.)

Step 7

Bodies

Step 8

Step 9

Step 8: Once again, let the bobbin hang down the off-side of the hook shank. Spin the dubbing fur onto the thread and tie on a dubbed fur body (refer back to the previous exercise, "Dubbed Fur Body," page 48, for directions). Tie the body over the tied-in tinsel. When you've half-hitched after tying in the body, let the bobbin hang.

Step 9: Pick up the long tinsel end in your *left* hand, take it behind the shank and down, transfer it to your *right* hand and bring it up underneath the shank and toward you. When the tinsel is on your side of the hook, angle it up and toward the eye of the hook, as pictured. This is your first wrap of tinsel.

DUBBED FUR BODY WITH TINSEL RIB

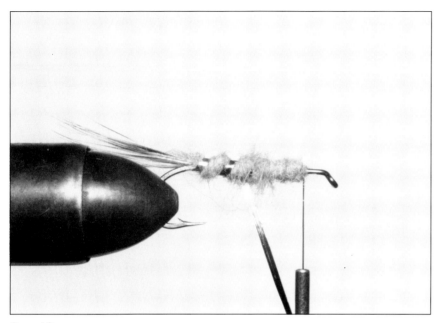

Step 10

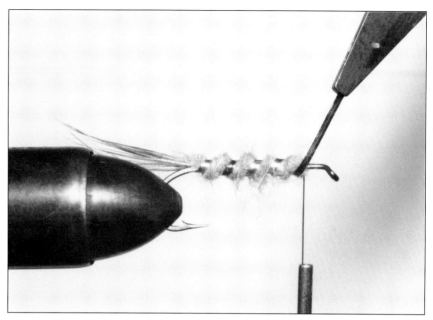

Step 11

Step 10: Continue to wrap the tinsel forward in this way: You should have the tip of the tinsel between the thumb and forefinger of your right hand. Take the tinsel over the top of the shank and down the off-side. When the tinsel is underneath the shank, take the tip between the thumb and forefinger of your *left* hand and bring it up toward you and to the top of the shank. Transfer the tip to your *right* hand again and repeat the procedure. (As you perform this method of wrapping you'll see very quickly why it's necessary: The hanging bobbin prevents you from wrapping exclusively with your right hand.) Space the tinsel wraps, or *ribs*, about one-eighth inch apart. Getting the spacing right on a tinsel body is the toughest part of tying one. Check the spacing after every wrap or so; if it isn't spaced correctly simply unwrap the tinsel and begin again.

Step 11: When you're satisfied with the spacing of the ribs, continue wrapping forward. Your final wrap of tinsel should be where the dubbed body ends, as pictured. (We're using hackle pliers in the photograph to hold the tinsel because it creates less visual distraction than do fingers. Use your fingers to wind tinsel.)

Bodies

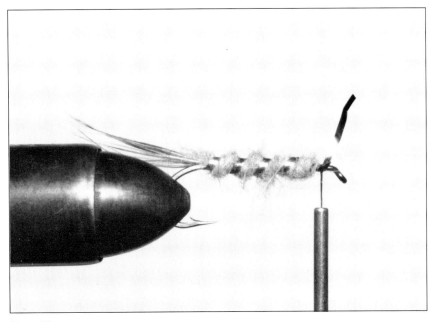

Step 12

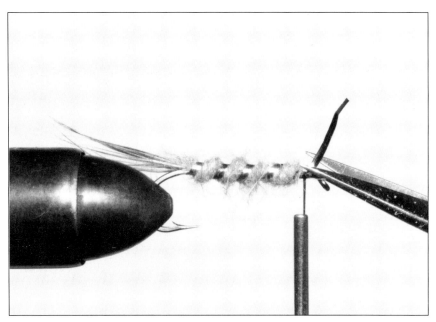

Step 13

Step 12: When you've wound the tinsel to the end of the body, you are ready to tie it off. Tie off in this way: Pull the tinsel end up with your *right* hand. With your *left* hand pick up the bobbin and make two to four wraps over the tinsel by dropping the bobbin over the top of the shank to the off-side, picking it up underneath the shank, and bringing it up and dropping it over the top again for one complete wrap. Snug these wraps against the body. Apply a half-hitch, pulling it under the tinsel end and snugging it against the body.

Step 13: Clip the excess tinsel end close to the body, being careful not to cut the thread.

DUBBED FUR BODY WITH TINSEL RIB

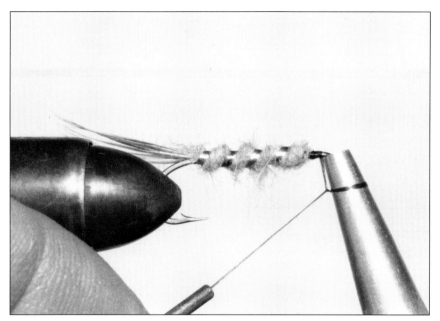

Step 14

Step 15

Step 14: Apply another half-hitch, pulling it next to the body.

Step 15: A side view of the completed dubbed fur body with tinsel rib. If you were following a fly pattern, you would go on at this point to add other components to the fly. Always be sure you have left room for them between the end of the body and the hook eye.

Step 1

Step 2

Chenille Body

For this exercise you will need a regular-length or 1XL wet-fly or nymph hook, #12 or #14. You will also need tying chenille in black, brown, or olive, and Monocord or nymph thread in black, brown, or olive (to match the color of the chenille).

Step 1: Secure the hook in the vise and tie in the thread. Cut a three- or four-inch length of chenille. Strip away the fibers on one end of the chenille to create a bare stem of about a quarter-inch in length.

Step 2: Tie in the chenille using the same procedure you learned in tying tail fibers and tinsel. When the chenille is in position on the top of the shank, apply a half-hitch to secure the material. It isn't necessary to trim the chenille end.

CHENILLE BODY

Step 3: Wrap the thread forward on the shank to a point about a quarter of an inch behind the eye of the hook, as pictured. There should be room behind the eye to tie off the chenille after it has been wrapped forward. (In the photograph we are holding the long end of the chenille up and back for a clearer view of the hook. It's not necessary for you to do this.) Let the bobbin hang.

Step 3

Step 4: Make an initial wrap of chenille by pulling it straight up, bringing it down the off-side of the hook, then under the shank and up toward you, ending with the chenille held straight up over the hook once again.

Step 4

Bodies

Step 5

Step 6

Step 5: Continue to wrap the chenille forward on the hook shank by taking the tip of the chenille in your *right* hand and moving it over the top of the shank and down the off-side of the hook. Take the chenille tip in your *left* hand when it's underneath the shank and bring it up toward you and to the top of the shank, where you'll once again take it in your right hand to repeat the wrap. Gently tighten each wrap and make sure it abuts the previous one smoothly and evenly. Wrap to the point where the thread wraps stop behind the eye of the hook.

Step 6: Tie off the chenille where the thread wraps end with two to four wraps of thread and a half-hitch. Pull the half-hitch over the eye of the hook and underneath the chenille end, tightening it against the body.

Basic Fly Tying

CHENILLE BODY

Step 7: Clip the excess chenille as close to where it's been tied off as possible. Take care not to cut the thread wraps.

Step 7

Step 8: A side view of the completed chenille body. We've wrapped a head on this body, which is a procedure you'll learn in a subsequent exercise. (If you were following a fly pattern, you would go on at this point to add other components to the fly.)

Step 8

Bodies

Peacock Herl Body

For this exercise you will construct a straight tail of your choice on a #14 or #16 regular length or 1XL hook, using Monocord or nymph thread in brown or olive. In addition, you will use peacock herl. This is available as whole peacock tail feathers, as the "eye" of a tail feather, or in packages containing herl fibers of approximately the same length.

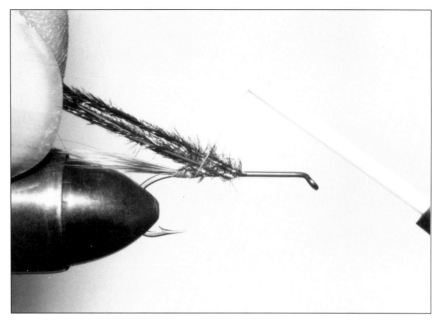

Step 2

Step 1: Secure the hook in the vise and tie on a straight tail.

Step 2: Select three peacock herl fibers; they should all be about the same length, from four to six inches. Combine the herl fibers into a bunch, place the bunch against the side of the hook, and tie it in with one loose wrap, using the same procedure you learned for tying in tail fibers and tinsel. The tie-in point for the herl bunch is the point on the hook where the tail-fiber ends are bound down by the thread. Pull the first wrap tight.

PEACOCK HERL BODY

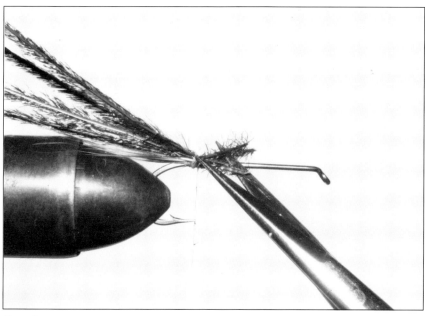

Step 3

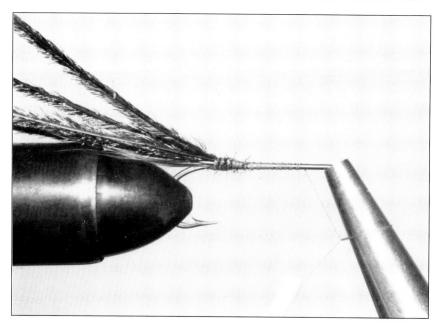

Step 4

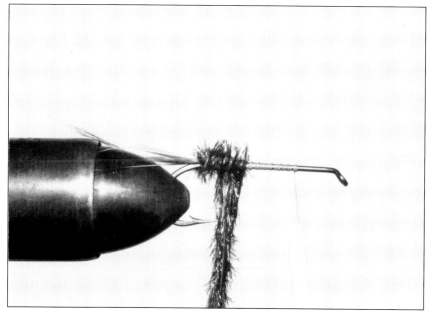

Step 5

Step 3: Make four or so more wraps to secure the peacock herl and bring it to the top of the hook shank. Pull each wrap snug. When the herl bunch is in position on the top of the shank, apply a half-hitch and clip off the excess ends.

Step 4: Begin to wrap the thread forward over the herl and continue forward on the hook shank to a point about three-quarters the length of the straight part of the shank. Apply a half-hitch and let the bobbin hang.

Step 5: Wrap the peacock herl forward toward the eye of the hook as follows: Grasp the tips of the herl bunch between the thumb and forefinger of your *right* hand and pull it down behind the hook shank. When the herl is underneath the shank, take the tips between the thumb and forefinger of your *left* hand and bring it up toward you and to the top of the shank. Transfer the herl tips to your *right* hand again and repeat the procedure. Make sure these wraps are tight and leave no gaps between the fibers.

Bodies

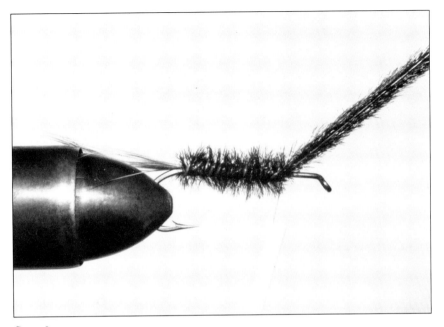

Step 6

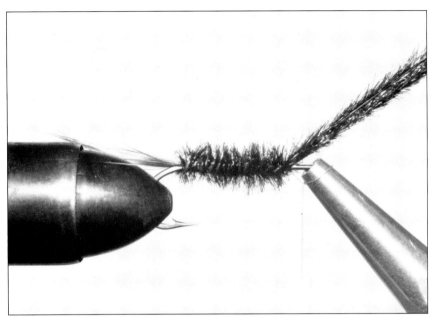

Step 7

Step 6: Continue to wrap the herl forward on the hook shank to the point where the thread wraps end.

Step 7: Tie off the herl with two to four wraps of thread and a half-hitch. Pull the half-hitch over the eye of the hook and underneath the herl, tightening it against the edge of the herl wraps.

PEACOCK HERL BODY

Step 8

Step 9

Step 8: Trim the excess herl as shown. It's not necessary to trim extremely close to the hook; just make sure the trimmed herl end blends with the wrapped herl body.

Step 9: A side view of the completed peacock herl body. (If you were following a fly pattern, you would go on at this point to add other components to the fly.)

Bodies

Clipped Deer-Hair Body

For this exercise you will need a patch of deer body hair, a #12 or #14 regular length hook, and nymph thread or Monocord in black or brown (to match the color of the deer hair). Regular 6/0 tying thread is not strong enough to flare the deer hair, which is the technique to be learned in this exercise.

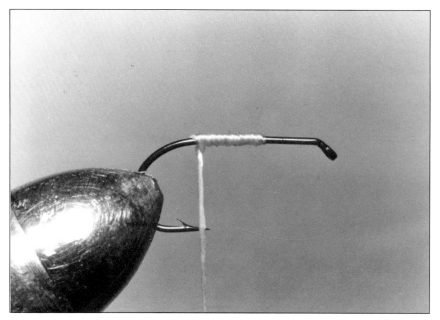

Step 1

Step 1: Secure the hook in the vise and start the thread just above the point of the hook. (For the photographs we've used heavy-duty stitching thread because it shows up better than tying thread does in the particular technique demonstrated here.) Wrap the thread forward to the midpoint of the hook shank, then back again over the forward wraps you've just made to the tie-in point. (Remember that when wrapping this close to the point of the hook you must wind the thread at an angle to keep it from being frayed or severed by the hook's sharp point.)

CLIPPED DEER-HAIR BODY

Step 2

Step 3

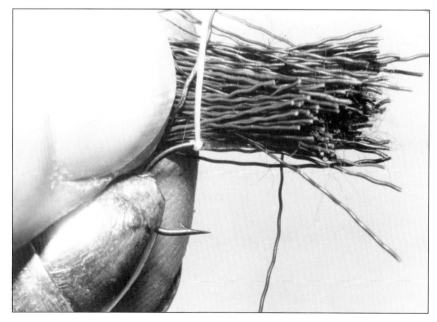

Step 4

Step 2: From the patch of deer hair, clip a bunch that is about the width of your thumbnail and about an eighth of an inch thick. Don't make the clump too thick; if you're in doubt as to the thickness, go with a thinner, rather than a thicker, volume.

Step 3: With the thumb and forefinger of your left hand, grasp the clump of hair at its midpoint, as pictured. Use your right hand to pick any fuzz out of the hair.

Step 4: Position the clump of deer hair on the top of the hook shank, with the butt ends toward the eye of the hook and with the midpoint of the clump at the last thread wrap at the *rear* of the shank.

Bodies

Step 5

Step 6

Step 5: Holding the deer hair with your left hand, pick up the bobbin in your right hand and begin one loose turn of thread, bringing the thread up toward you, over the top of the hair clump, and down the off-side of the hook shank.

Step 6: Pull the bobbin firmly so that the butts of the deer hair begin to flare slightly, as pictured. Continue to hold the deer hair with your left hand.

Step 7: Make a second turn of thread, in front of and abutting the first one, and pull it tight. The deer hair will begin to flare even more, as pictured. Let go of the hair clump with your left hand.

Step 8: Make a third turn of thread and pull it tight. Make a maximum of four more turns, wrapping forward and pulling each turn snug. The deer hair should flare a little more with each wrap. When the hair is flared to the extent shown in the photograph, you've wrapped enough. (It is difficult to see in a photograph where the thread wraps are. As you tie, however, you will discover that it is easier than it looks here to tie within the flared hair and that the thread wraps create a small flat space inside the hair.) Let the bobbin hang freely.

CLIPPED DEER-HAIR BODY

Step 7

Step 8

Step 9: Clip off another clump of deer hair of the same proportions as the first one. Position this second clump on the hook shank just in front of the first clump. You'll have to snug the second clump against the first, but do this gently—you don't want to create a great deal of tension. As you did for the first clump of deer hair, make one loose wrap of thread and pull it tight so that the hair begins to flare slightly. Make this first wrap by picking up the bobbin and bringing the thread up from underneath the shank toward you, then up and over the top of the second clump of deer hair, circling around the midpoint of the second clump.

Step 9

Bodies

Step 10

Step 11

Step 10: Continue wrapping forward through the second clump of deer hair, pulling the thread tight and flaring the hair a little more with each wrap. When the hair is flared as pictured, you've wrapped enough (maximum six to eight wraps). If more deer hair is needed to cover the hook, tie in additional clumps in the same manner.

Step 11: Apply a half-hitch, placing the end of the half-hitch tool over the eye of the hook and gently pushing the tips of the deer hair back and away in order to snug the half-hitch over the hook shank and up against the last wraps in the flared hair. Repeat with two more half-hitches. Whip-finish and clip the thread off close to the hook shank, being careful not to cut into the deer hair or the previous thread wraps.

CLIPPED DEER-HAIR BODY

Step 12

Step 13

Step 14

Step 12: Now you will begin to trim the flared deer hair to fashion a body. Position your heaviest pair of tying scissors down on top of the hook shank, as close as you can get to the shank through the hair. Clip the hair of the first one-third of the body length.

Step 13: Continue clipping the top of the fly in this way to the end of the shank at the rear of the hook. The hair should be as flat on the top of the hook as you can make it. (Pictured is a side view of the completely clipped top.)

Step 14: Repeat the clipping procedure on the bottom of the hook shank, but position your scissors midway between the barb of the hook and the underside of the shank.

Bodies

Step 15

Step 16

Step 15: Pictured is a side view of the deer-hair body clipped flat on top and bottom.

Step 16: We've switched to a top view of the fly in progress to better demonstrate the next two steps. In the same manner as before, clip the hair on one side of the hook shank.

CLIPPED DEER-HAIR BODY

Step 17

Step 18

Step 17: Clip the hair on the other side of the hook shank.

Step 18: Pictured is a top view of the finished clipped deer-hair body.

Step 19: A side view of the finished clipped deer-hair body. This is the first exercise that has resulted in a fly you can actually fish with. Fish this generalized deer-hair beetle dry.

Step 19

Bodies

Quill Body

The quill body you will tie in this exercise is made from a chicken-hackle quill. You will construct a straight tail of your choice on a #12 or #14 regular length hook, using either 6/0 thread, Monocord, or nymph thread in black, brown, or olive. (Normally, the thread used depends on the size and pattern.) In addition, you will need a hackle from either a wet- or dry-fly neck.

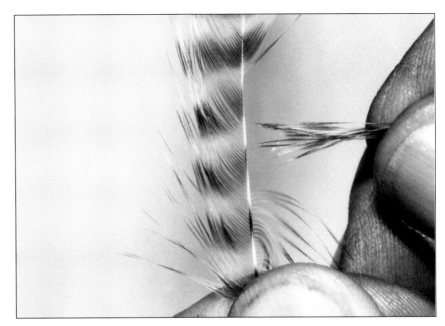

Step 1

Step 1: Select the longest hackle you can get from the neck. Leaving a few fibers on the tip end of the quill, strip all the rest of the fibers from the quill by gently pulling them off. Soak the quill in a glass of tepid water for about a minute. This will make the quill more supple and easier to wrap. It will also help prevent splitting, particularly if the quill is old (and unless you're a professional fly tier, your necks may well sit around for years, until you've tied enough flies to use all the hackles).

QUILL BODY

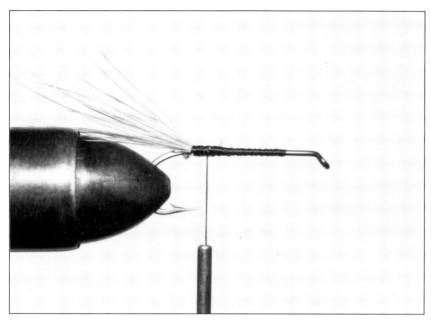

Step 2

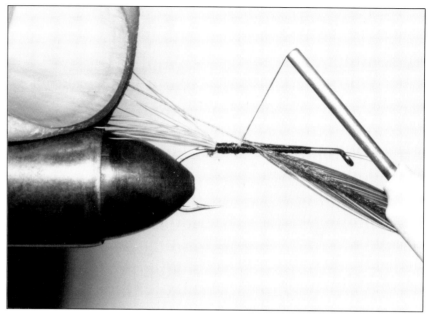

Step 3

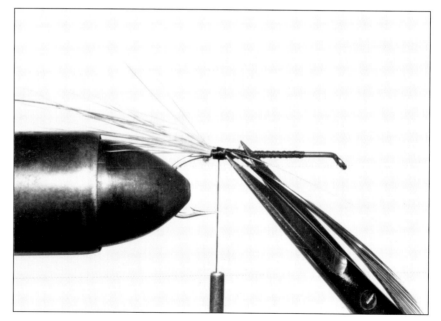

Step 4

Step 2: Secure the hook in the vise and tie on a straight tail. Wrap the thread forward on the shank about three-quarters of the way to the eye, then wrap back over these wraps to the point at which the tail is tied in. This wrapping will provide an underbody for the quill wraps to come. It is important to allow a sufficient distance (about a quarter of an inch or so) between the end of the thread wraps and the eye of the hook for a thorax, wings, hackle, and/or head, depending on the requirements of the pattern followed.

Step 3: Tie in the hackle quill near its narrow end, where you've left some fibers attached, by making two to three wraps of thread. It's not necessary to wrap until the quill sits on top of the shank. Let the bobbin hang down the off-side of the shank.

Step 4: Clip off the excess hackle tip, being careful not to cut into the thread.

Bodies

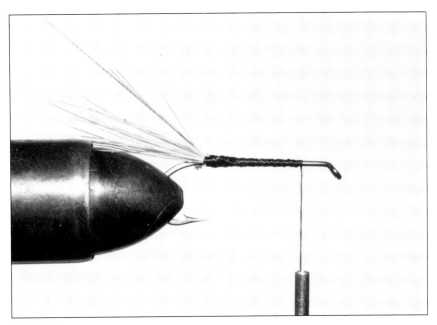

Step 5

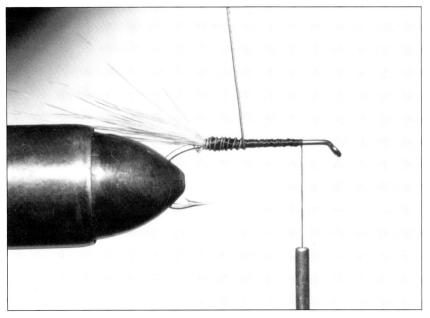

Step 6

Step 5: Wrap the thread forward over the cut end of the quill to the point where the thread underbody ends at the front of the shank. Apply a half-hitch.

Step 6: Using your fingers, begin to wrap the quill forward on the hook shank, making certain each wrap is pulled snug and abuts the previous wrap smoothly. (Because quills have a light and dark side, the quill wraps in the photograph look as if there is a space between them. There isn't.)

QUILL BODY

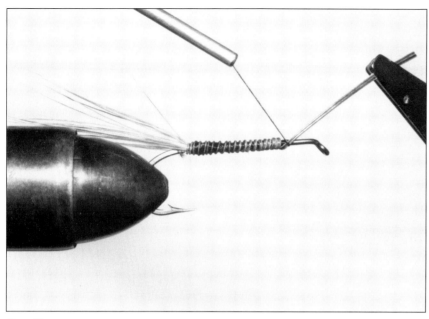

Step 7

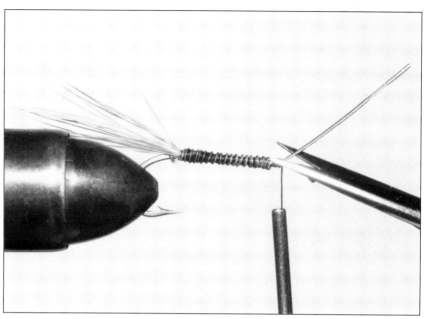

Step 8

Step 7: Continue wrapping the quill forward to the point where the thread underbody stops. Tie off the quill with two to four wraps of thread and a half-hitch.

Step 8: Clip off the excess quill as close to the hook as you can, being careful not to cut into the thread.

Step 9: A side view of the completed quill body. You will use this body, as is, for the next exercise. Leave it in the vise if you wish to go on at this time.

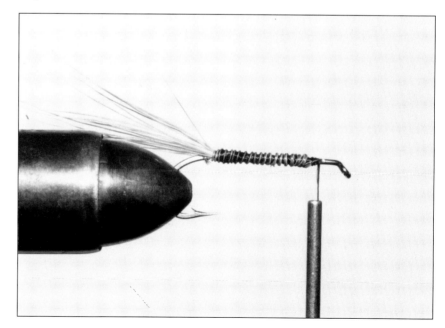

Step 9

Bodies

Thorax and Wing Case

A number of patterns require the front end of a fly to be a larger diameter than the other parts of the body. This enlarged part of the imitation is called the *thorax* and corresponds to the area immediately behind the head of the natural insect. In many nymph imitations the thorax serves as a base for wing cases.

This exercise uses the quill body you tied in the previous exercise as the starting point for constructing a dubbed-fur thorax, wing case, and throat, as called for in many nymph patterns. In addition to the quill body, you will need gray dubbing fur and a duck wing feather. Duck wing feathers come packaged as a set of two feathers, one from the right wing and one from the left wing of the same bird.

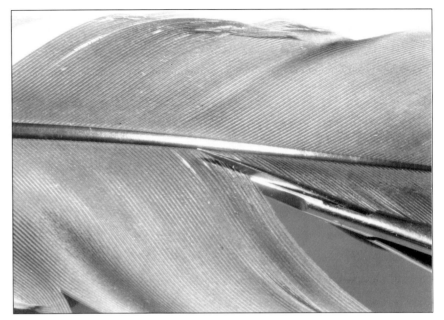

Step 1

Step 1: From a duck wing feather, cut a section about one-half inch wide. Be careful not to separate the barbs from the cut section, and make your cut with fine scissors fairly close to the quill, as pictured.

THORAX AND WING CASE

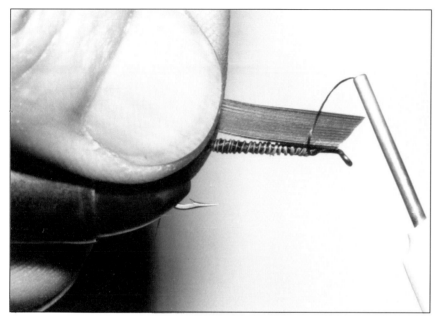

Step 2

Step 3

Step 2: Fold the cut wing section in half lengthwise. Hold the folded section flat on top of the hook shank, with the trimmed edge toward the eye of the hook. Make one loose turn of thread and pull it tight.

Step 3: Make four to six wraps of thread forward over the wing section, tying it in. Apply a half-hitch, pull it snug under the wing section, and clip off the excess wing section close to the thread wraps.

Step 4: Wrap forward just enough to cover the cut end of the wing section, then back again an equal number of wraps. Prepare a small amount of dubbing fur to spin on the thread.

Step 4

Bodies

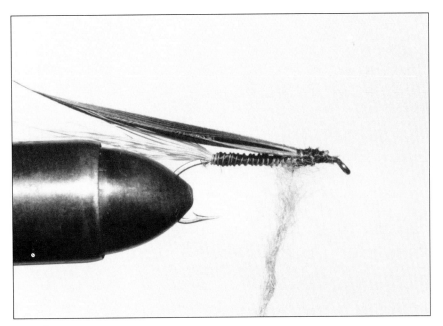

Step 5

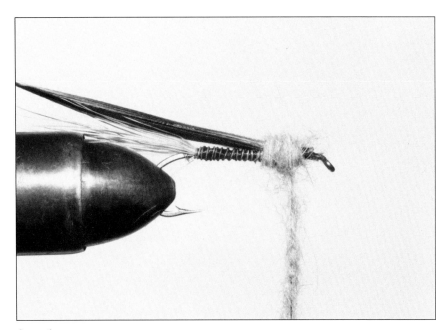

Step 6

Step 5: Spin the dubbing on the thread.

Step 6: Wrap the dubbed thread forward on the hook toward the eye, leaving enough room to tie a head—about an eighth of an inch or so.

Basic Fly Tying

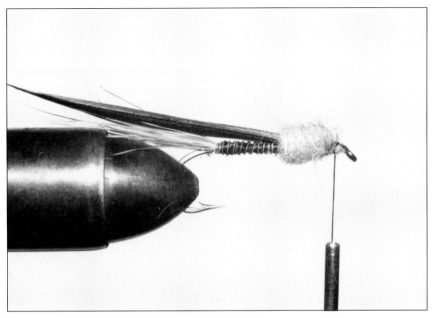

Step 7

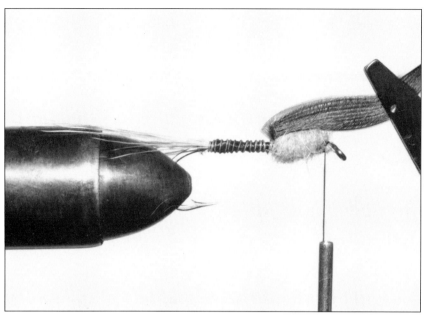

Step 8

Step 7: Wrap the dubbed thread back over the previous wraps to the point at which you began wrapping, then forward once again toward the eye of the hook. As you can see from the photograph, this builds a fat ball of fur to represent the thorax of an insect. Apply a half-hitch and snug it against the fur thorax. (If your thread is still covered with dubbing at the point you must make the half-hitch, merely pull the fur from the thread before tying the knot.)

Step 8: Use your right hand to pull the tied-in duck wing section forward over the thorax toward the eye of the hook, as pictured. (Again, to create a more uncluttered image, we're using hackle pliers instead of fingers in the photograph.)

Bodies

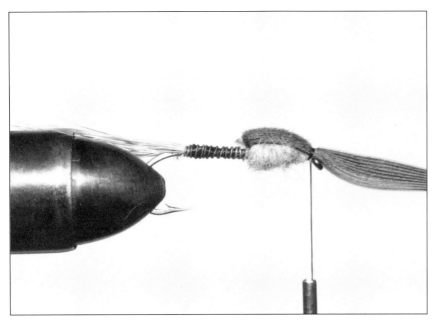

Step 9

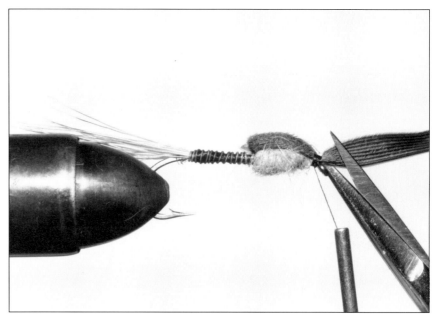

Step 10

Step 9: Continue to hold the duck wing section forward and flat over the thorax. Tie off the wing section just behind the eye of the hook by making three or four wraps of thread in this manner: Pick up the bobbin in your left hand, bring it up from underneath the hook shank toward you, continue moving the bobbin over the hook shank. When you've reached the top of the shank, just drop the bobbin down the off-side of the shank, pick it up again when it's underneath the shank, and repeat the process. (This may feel awkward the first few times you do it, and you will need to be extra careful to make these wraps snug.) Apply a half-hitch, pull the knot over the eye of the hook and under the wing section, and snug it against the thorax.

Step 10: Clip off the excess wing section close to the wraps, being careful not to cut the thread.

THORAX AND WING CASE

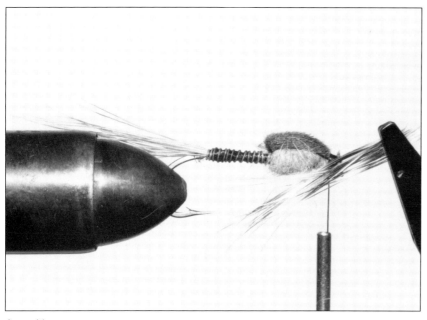

Step 11

Step 12

Step 11: Select a bundle of hackle fibers from a wet-fly hackle for the throat of the fly. The fibers should be of the same thickness as if you were using them to make a straight tail, and their finished length should be as long as the straight portion of the hook shank.

Step 12: Position the fibers *below* the hook shank (not on the side) at the angle pictured. Tie in the fibers, holding them on the bottom of the shank with your *right* hand, with three or four thread wraps taken with your *left* hand (use the "bobbin dropping" maneuver you learned in Step 9). Apply a half-hitch and clip the excess fibers.

Step 13: A side view of the completed thorax, wing case, and throat on the quill body. We've wrapped a head on this body, which is a procedure you'll learn in a subsequent exercise.

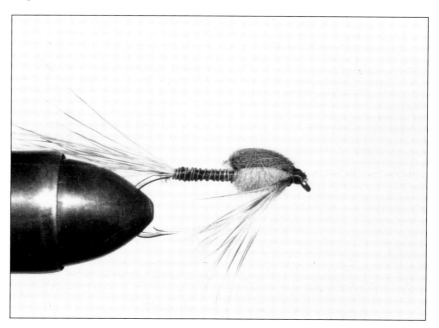

Step 13

Bodies

Foam Body

This exercise involves tying a simple but complete dry fly: a black ant. You will use a regular dry-fly hook in any size from #16 to #20. This is a smaller hook than you have used yet, and you may find the size difference glaringly apparent. Don't let it throw you; slow down if you have to, but stick with it—you'll soon get used to tying on smaller hooks. In addition, you will need regular 6/0 tying thread in black, black hackle from a dry-fly neck, and a black closed-cell foam cylinder. Closed-cell foam for fly tying comes in sheets to be cut as needed, and in packets of preformed cylinders. The cylinders are quite convenient, and we recommend them for tying foam bodies.

We're jumping the gun a bit here by applying hackle to complete the black ant. But since there are no tails, wings, or other complications involved in this pattern, it's worth a try. If you're uncomfortable with the procedure, refer to the sequence on "Winding Dry-Fly Hackle" on page 113 for more detailed instructions and photographs.

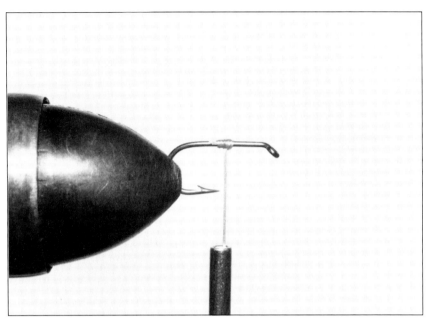

Step 1

Step 1: Secure the hook in the vise and tie the thread on in the center of the hook. (Once again we are using heavy-duty stitching thread for a clearer image in the photographs.)

Basic Fly Tying

FOAM BODY

Step 2: Pull a black foam cylinder from the packet (they come in three-inch lengths) and tie it on top of the hook shank with two or three firm wraps. The front end of the cylinder should end just above the eye of the hook, as pictured.

Step 2

Step 3: Clip off the rear end of the foam so the body ends just above the bend of the hook shank.

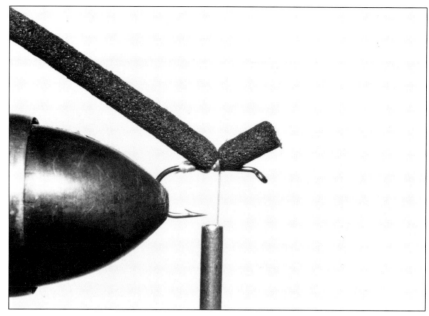

Step 3

Bodies

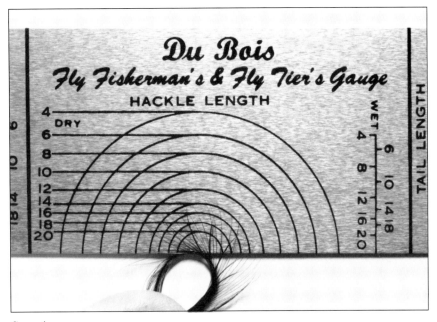

Step 4

Step 5

Step 4: Use your hackle gauge to select the proper size hackle (a #14 hook takes size 14 hackle, a #16 hook takes size 16 hackle, and so on). The gauge makes this a simple procedure. Grasp the butt of the hackle you've chosen between the thumb and forefinger of your right hand. With your left hand, grasp the tip of the hackle and bend it in a circle. Take the feather circle in your right hand. Lay the hackle fibers from the circled feather against the measuring lines on the gauge, making sure the bent quill rests against the edge of the gauge or whatever starting point is indicated on your particular gauge. The hackle in the photograph is a size 14 because its fiber tips reach the line indicating size 14.

Step 5: Strip the webby fibers from the base of the feather, then tie the hackle, butt end forward, on the side of the hook shank in the middle of the foam body with three or four thread wraps. Clip the hackle end close to the hook.

FOAM BODY

Step 6: Take the tip of the hackle between the thumb and forefinger of your right hand (or you can use hackle pliers) and wind it forward three full turns. Make sure to snug each wrap and wind smoothly so that each turn abuts the previous one. Tie off the hackle with two to four wraps of thread and a half-hitch. Clip off the excess hackle close to the foam body. Wind the thread forward (it should be positioned under the foam body on the hook shank after the half-hitch) all the way to the eye of the hook. Whip-finish and clip off the thread as close to the hook as you can.

Step 6

Step 7: A side view of the finished foam-body ant. This is another fly you can fish with, and it's a very effective pattern.

Step 7

Bodies

Wings

Part 5

- Upright Wood-Duck Wings
- Poly Wings, Spent
- Hackle-Tip Wings
- Tent-Type Wings

Most dry flies are tied to imitate the mayfly in either its dun or spinner stage. When the emerged insect (dun) floats on the surface before taking flight, the wings are its most visible feature; the uprightness of the wings reminds many people of tiny sailboats. Hackle tips, cut to shape or "as is," and hackle or hackle-like fibers are commonly used for upright wings. In this section you will learn to tie upright wings with fibers stripped from the flank feathers of the wood duck and with chicken-hackle tips. Many other feathers, hairs, and synthetic materials may be used to tie upright wings, and the techniques you learn in the exercise "Upright Wood-Duck Wings" will stand you in good stead when following patterns requiring hair or synthetic wings.

After the female mayfly returns to the water to deposit her eggs (this is the spinner stage), she floats helplessly on the surface and her wings are no longer upright but *spent*, immersed in the surface film. The use of polypropylene yarn has become increasingly popular for spent-wing imitations. Although the ease with which poly yarn may be used in tying probably accounts for much of its popularity as spent wings, its buoyancy and quick-drying properties add to its effectiveness. In the exercise "Poly Wing, Spent," you'll learn an important thread wrap, the figure eight, that is used again and again in fly tying.

Many wet-fly patterns require the wings to be tied over the body in a draping, tentlike manner; these are often referred to as *down-wing* patterns. Tent wings are usually cut from duck, turkey, or grouse wing feathers; wing feathers from many other birds will work, as well.

Many tent-wing patterns specify that the two wings be cut from two feathers, one of which comes from the right wing of a specific kind of bird, and the other of which comes from the

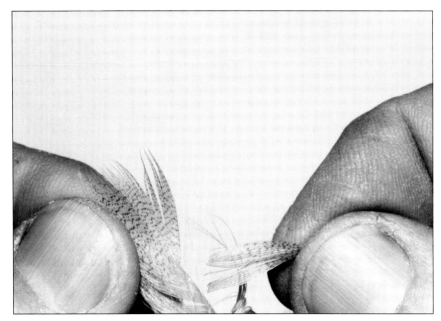

Step 2

left wing of the same bird. As with many techniques in fly tying, there is not universal acceptance of this as an absolute requirement. We think that, although such matched wings may be more pleasing esthetically, they do little, if anything, to enhance success on the stream. The bottom line is: If you're tying for competition, use matching left and right wing feathers. Otherwise, use the technique demonstrated in the exercise "Tent-Type Wings."

Upright Wood-Duck Wings

For this exercise you will need a #12 or #14 regular length dry-fly hook, regular 6/0 tying thread, and a wood-duck flank feather. (These feathers come packaged by the dozen.)

Step 1: Secure the hook in the vise and tie in the thread about a quarter of an inch behind the eye of the hook. (Because this exercise demonstrates a technique for tying in wings, the tie-in position is toward the front of the hook—where your thread would be after you had tied a body and were ready to attach wings.)

Step 2: Select a wood-duck feather. Hold the feather by its stem between the thumb and forefinger of your left hand and grasp a one-inch-wide section of fibers on one side of the quill with the thumb and forefinger of your right hand. Gently strip the fibers from the quill (Don't use the webby and fuzzy fibers toward the base of the feather.)

Step 3: Hold the same feather's stem between the thumb and forefinger of your *right* hand and strip an equal amount of fibers from the same location on the quill, but on the *opposite side*.

Step 3

Step 4

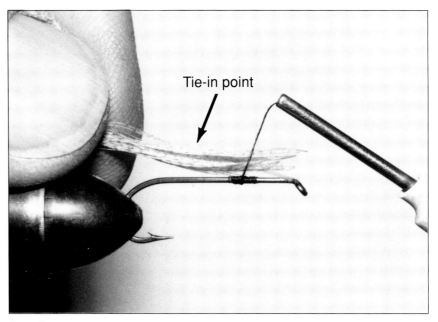

Step 5

Step 4: Combine the two fiber sections into one bunch with their tips even, as pictured.

Step 5: Position the fibers on top of the hook shank, stem-ends facing the eye of the hook, and measure them against the straight portion of the shank for proper length. The wings when finished and trimmed should be as long as the straight portion of the shank. Make a mental note of the correct length before you tie the fibers down. Make one loose turn of thread at the point you've noted on the fibers as the correct wing length and pull it tight.

Step 6: Make four to six thread wraps forward. Pull each wrap tight and make sure it abuts the previous wrap smoothly. Apply a half-hitch, snugging it under the fiber stems and against the wraps.

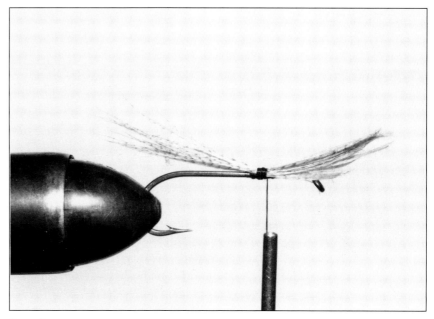
Step 6

Step 7: Clip off the excess fiber ends close to the hook.

Step 7

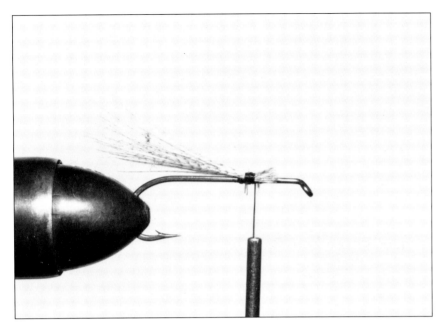

Step 8

Step 9

Step 8: A side view of the fiber wings tied down with ends trimmed. Because you'll be wrapping over the trimmed ends, it's not necessary to trim them extremely close to the hook.

Step 9: Wrap the thread forward to cover the trimmed fiber ends, then wrap backward to the end of the previous wraps, as shown. Let the bobbin hang.

UPRIGHT WOOD-DUCK WINGS

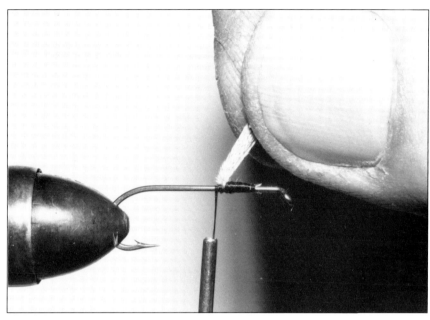

Step 10

Step 11

Step 10: Grasp the fiber-wing tips between the thumb and forefinger of your right hand and pull the wing forward toward the eye of the hook, as pictured. Hold the wing forward like this and pick up the bobbin with your left hand. Make four turns of thread close behind the wing as it is held forward by using the "drop bobbin" technique you learned in the "Thorax and Wing Case" exercise. Be sure these turns are pulled snug and are made right up against the wing fibers.

Step 11: Make four to six additional turns of thread behind the wing, still holding it forward with your right hand. Build a small hump of thread that will hold the wing in an upright position, as shown. To do this, wrap two or three times backward toward the hook bend, then forward toward the wing fibers two or three wraps. When your wings stand upright as pictured, bring the thread forward to make one complete wrap in front of the wing and apply a half-hitch. The upright wing is complete.

Wings

Poly Wings, Spent

For this exercise you will need a #12 or #14 regular length dry-fly hook, poly yarn in cream, tan, or white, and regular 6/0 tying thread in white. We are using heavy-duty stitching thread and black poly yarn in the photographs for effective contrast, not realistic color. And in order to best illustrate the figure-eight wrap, an important technique in tying wings, all the photographs in this exercise are top views.

Step 2

Step 1: Cut a two-inch length of poly yarn to be used for the wings. With small flies and thick yarns, you might have to separate the strand of yarn into two or more strands. Secure the hook in the vise and tie in the thread about a quarter of an inch behind the eye (remember, this is the position at which you would tie in wings if following a pattern).

Step 2: Hold the two-inch piece of poly yarn in your left hand, between thumb and forefinger. Lay the yarn on top of the shank and across it at the slight angle shown in the photograph. Hold the yarn on top of the hook while you begin the figure-eight wrap, as follows: Begin one loose turn of thread by taking the bobbin in your right hand and bringing it up toward you on a slight diagonal so that the thread comes up *behind* the poly wing on the side of the hook shank that faces you. Continue to move the bobbin over the hook shank on the same diagonal arc, taking the thread over the poly wing and *in front of it*, then down the off-side of the shank. Pull this first wrap tight.

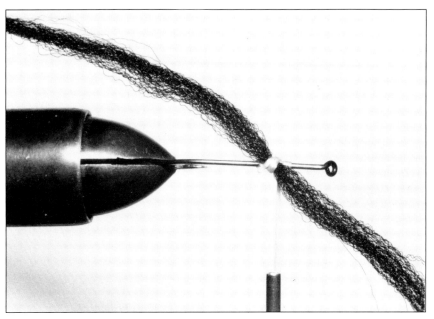

Step 3

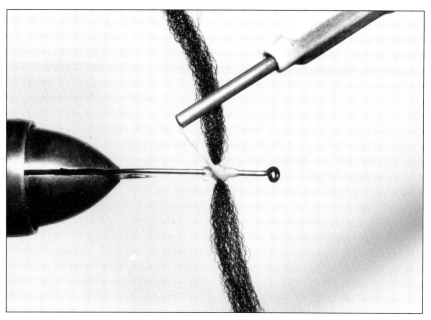

Step 4

Step 3: Make three additional wraps in the same manner as the first, wrapping behind the wing on your side of the shank and over and in front of the wing on the off-side of the shank. Each additional wrap is made on top of the previous one. Be sure to snug each wrap tight.

Step 4: Make one loose wrap in a reverse manner: Bring the thread up *in front of* the poly wing on your side of the shank, take it backward on a diagonal over the wing and bring it down *behind* the poly wing on the off-side of the shank. Pull this wrap tight.

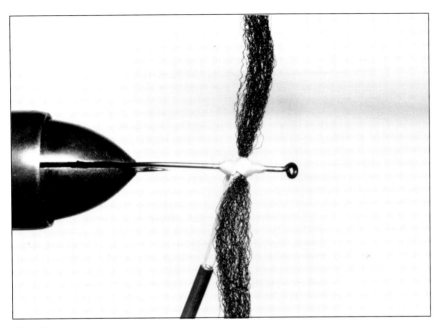

Step 5

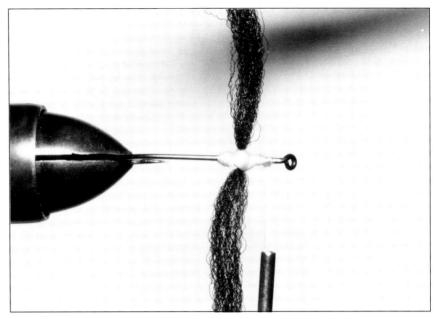

Step 6

Step 5: Make three more wraps like this, wrapping in front of the wing on your side of the shank and bringing the thread over and down behind the wing on the off-side of the shank. Make each wrap on top of the previous one and pull it tight. You've now completed the figure-eight wrap. The poly wing should be relatively straight across the shank because each opposing diagonal wrap works to pull the wing in a straighter line.

Step 6: Wrap the thread once toward the eye of the hook by completing one full turn on the hook shank in front of the wing. Apply a half-hitch. (Your thread wraps will not be, and should not be, as bulky as those in the photographs. The heavy thread we're using is for illustrative purposes only; regular 6/0 thread would not present as clear a picture of the figure-eight wrap.)

SPENT POLY WINGS

Step 7

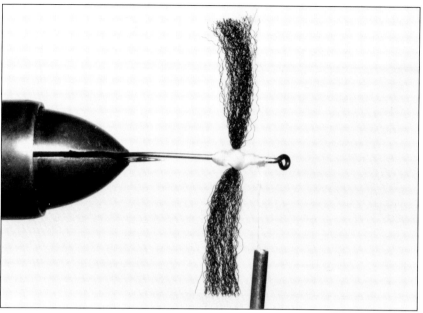

Step 8

Step 7: Clip the poly wings to correct length. Each should be as long as the straight portion of the hook shank. You may find your hackle gauge or a ruler useful in measuring poly wings for correct length.

Step 8: A top view of the completed spent poly wings with figure-eighted thread.

Wings

Hackle-Tip Wings

For this exercise you will need a #12 or #14 regular length dry-fly hook, hackles from a dry-fly neck (we're using grizzly in the photographs), and regular 6/0 tying thread.

Step 2

Step 1: Secure the hook in the vise and tie in the thread about a quarter of an inch behind the eye of the hook.

Step 2: Select two hackles from the dry-fly neck—the hackles should be the same length. Measure the hackles for proper length; the finished wings should be as long as the straight portion of the hook shank. Strip some fibers below the point you have noted as the correct wing length on each hackle to expose a half-inch or so of bare stem.

HACKLE-TIP WINGS

Step 3: Hold the hackle tips between the thumb and forefinger of your left hand and lay the bare stems against the side of the hook shank facing you at the thread tie-in point. The tips of the hackle should face backward, over the hook bend. Make one loose turn of thread around the stems and pull it tight.

Step 3

Step 4: Make four to six turns of thread forward over the stems, until they are in position on top of the hook shank. Apply a half-hitch, snugging it under the stems and up against the thread wraps. Clip off the excess stems.

Step 4

Wings

Step 5

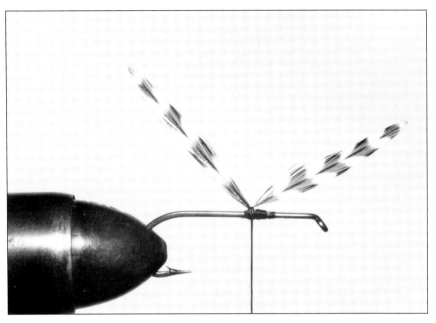

Step 6

Step 5: Grasp the hackle tips between the thumb and forefinger of your right hand and pull them forward, toward the eye of the hook, as you did in the "Upright Wood-Duck Wings" exercise. Make two thread wraps immediately behind the hackle-tip wings, pulling each tight. Make four to six more wraps to build up the slight hump of thread behind the wings that will hold them upright. Bring the thread in front of the wings, make one complete wrap there, and apply a half-hitch.

Step 6: Begin a figure-eight wrap by taking one loose turn of thread as follows: Bring the thread up toward you on a slight diagonal and wrap *behind* the wing on your side of the shank, continue on the same diagonal arc by wrapping forward over the shank in between the two hackle tips, and bring the thread down *in front of* the wing on the off-side of the shank. Pull this wrap tight. Make three to five additional wraps in the same way and pull each snug. Now reverse the direction of the wrap: Begin one loose turn of thread by bringing it up toward you *in front of* the wing on your side of the shank, continue the wrap backward over the shank in between the hackle tips, and bring it down *in back of* the wing on the off-side of the shank. Pull the wrap tight and make three to five additional wraps in the same way. You've completed the figure-eight wrap, which will pull the wings down slightly.

HACKLE-TIP WINGS

Step 7: Bring the thread forward, in front of the wings, and complete four to six forward wraps. Apply a half-hitch. Pictured is a side view of the hackle-tip wings tied on in proper position.

Step 7

Step 8: A front view of the hackle-tip wings properly divided by the figure-eight wrap.

Step 8

Tent-Type Wings

In order to illustrate this type of wing as it would be tied on a wet fly, we've used a duck-wing feather and attached it to the peacock herl body with straight hackle-fiber tails you learned to tie in Part Four (page 62). For this exercise you will tie that body according to those instructions. In addition, you will need a mallard wing feather and head cement.

Step 1

Step 1: From the mallard wing feather, cut a section half an inch wide from one side of the stem.

TENT-TYPE WINGS

Step 2

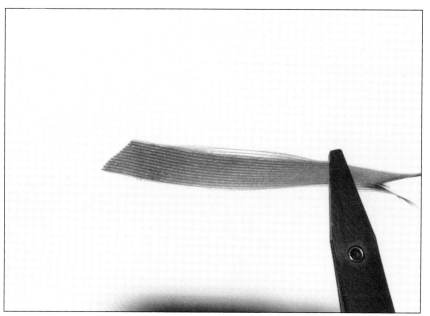

Step 3

Step 2: The section of wing you have cut should look like this. Hold the section, as pictured, between the thumb and forefinger of your left hand.

Step 3: Using your right hand, fold the wing section in half lengthwise. (We're using hackle pliers in the photograph because they are not as visually distracting as fingers.)

Step 4: Take the cut end of the folded wing section between the thumb and forefinger of your *right* hand, with the tip end extending back over the hook, and measure the wing against the hook shank for proper length. The tip of the finished wing should end just above the bend of the hook. Make a mental note of the proper length so you will know where to tie in the wing section.

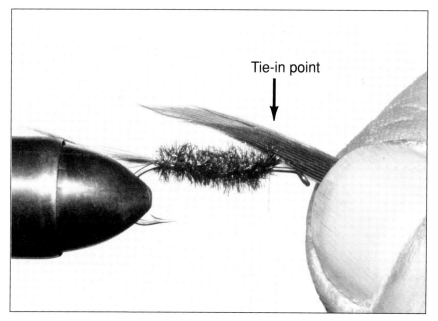

Step 4

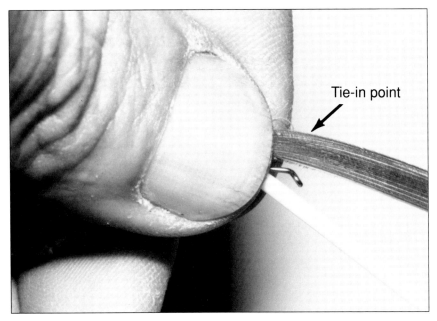

Step 5

Step 6

Step 5: Grasp the folded wing section between the thumb and forefinger of your *left* hand. Position the wing on top of the hook shank at the point you've noted as the location to tie in the wing, as pictured. Firmly grasp the wing *and* the hook shank, and with your right hand begin one loose thread wrap by bringing the thread up toward you from underneath the shank, over the wing and shank, then down the off-side, leaving a loose loop above the wing, as shown. Pinch the wing, hook shank, *and* loose loop of thread between the thumb and forefinger of your left hand and pull the loop tight, letting it slip between your thumb and finger straight down onto the wing section. This is a tricky maneuver because you shouldn't be letting the wing and hook shank go at any time.

Step 6: Continue to hold the wing, hook shank, and thread wrap and make three to four additional wraps forward on the wing section. Pull each wrap snug. When the wing is securely tied in you can let go with your left hand.

Step 7

Step 8

Step 7: Apply a half-hitch, snugging it under the excess wing fibers and up against the body.

Step 8: Clip off the excess wing-section fibers as close to the thread wraps as you can, being careful not to cut into the thread.

Step 9

Step 10

Step 9: A side view of the wing tied in, half-hitched, and with the wing-section fibers clipped.

Step 10: Now you are ready to construct a smooth tapered head on the fly. Do this by wrapping forward over the cut fiber ends toward the eye of the hook, as pictured, then back again until you've made a head of tying thread that slopes smoothly toward the hook's eye. There is no specific number of wraps needed. Rather than count all the wraps, learn to recognize what a proper head looks like so that you can stop wrapping when you've achieved it.

Basic Fly Tying

Step 11

Step 12

Step 11: Whip-finish the head of the fly.

Step 12: Clip the thread closely and apply a drop of head cement to the head of the fly.

Step 13: A side view of the completed fly.

Step 13

Hackle

Part 6

- Winding Dry-Fly Hackle
- Wet-Fly Hackle
- Palmered Hackle

Hackle wrapped around a fly body so that it radiates in a circle imitates the legs, wings, or breathing appendages of various insects. Wound hackle enhances the silhouette and float of dry flies and gives lifelike motion to wet flies and nymphs. Flies may be hackled lightly or heavily, may incorporate hackle as a *collar*, or may have *palmered* hackled that is sparsely wound the entire body's length.

The following exercises demonstrate the conventional methods of winding dry-fly hackle and wet-fly hackle, and we recommend you experiment by using these methods to wind sparsely and heavily. Hackling a dry fly is one operation where you will appreciate the quality of a grade A or 1 dry-fly neck; there should be minimal webbing on a properly hackled dry fly.

The last exercise in this section, and the book, will teach you how to palmer-hackle a Woolly Worm.

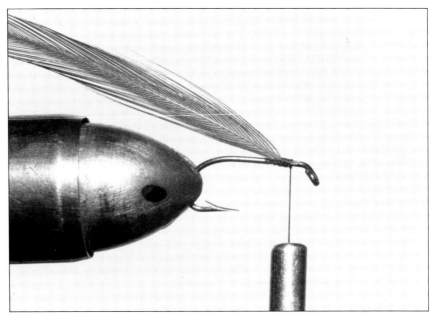
Step 1

Step 2

Winding Dry-Fly Hackle

This exercise gives you the opportunity to practice winding hackle as you would when following a dry-fly pattern that calls for wound hackle. Use hackle from a dry fly neck of any color, a regular length dry-fly hook sized to match the hackle, and regular 6/0 tying thread.

Step 1: Select a hackle from a dry-fly neck and measure it on your hackle gauge in order to size it (the hackle you use should be about the same size as the shank of the hook it will be tied on). Strip the webby fibers from the bottom third of the hackle's stem. Secure the hook in the vise and tie the thread in about a quarter of an inch behind the eye of the hook. (In most fly patterns the hackle is tied on in front of the wings.) Tie in the hackle on the side of the hook shank that faces you with four to six turns of thread, wrapping until the stem is in position on top of the shank. Apply a half-hitch and clip the excess hackle stem. Allow the bobbin to hang.

Step 2: The rest of the photographs in this sequence are front views, looking straight at the eye of the hook. Using hackle pliers, grasp the tip of the hackle and wrap forward as much as necessary to wind the remaining length of bare stem around the hook shank.

Step 3: Proceed to wrap the hackle forward on the hook shank, making certain the fibers remain perpendicular to the shank and that each wrap abuts the previous one smoothly and snugly. The photographs that follow illustrate how the hackle spreads around the hook at each quarter turn for a total of one and three-quarters turns.

Step 4: Front view of the second quarter-turn.

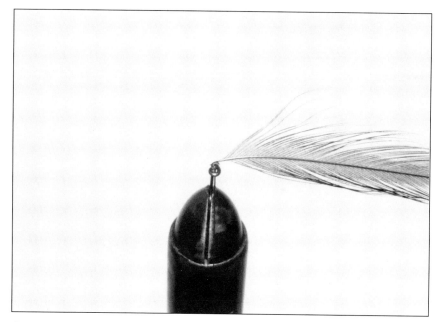
Step 3

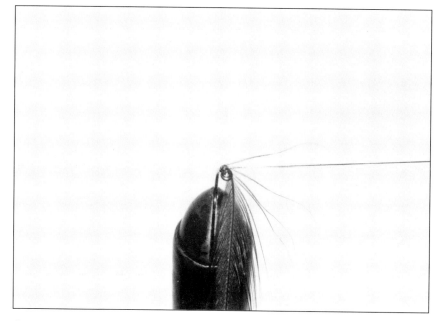
Step 4

Hackle

Step 5

Step 6

Step 5: Front view of the third quarter-turn.

Step 6: Front view of the fourth quarter-turn and first full wrap. Notice how the hackle is beginning to spread in a circle around the shank, as if it were a collar.

Basic Fly Tying

DRY-FLY HACKLE

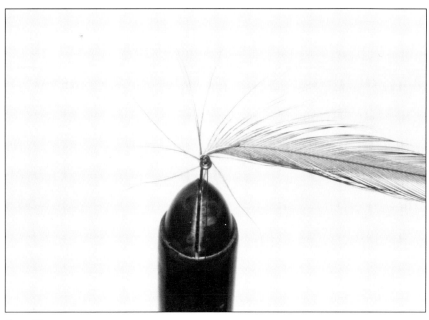
Step 7

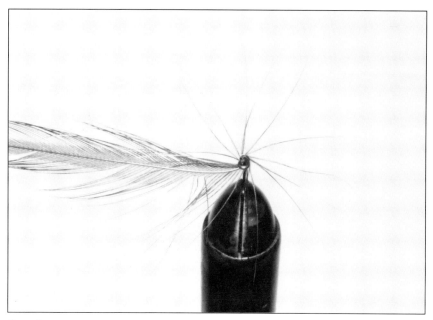
Step 8

Step 7: Front view of the fifth quarter-turn.

Step 8: Front view of the six and seventh quarter-turns.

Step 9: Continue to wrap the hackle until the desired proportion is achieved: this is usually six or eight complete turns for a standard dry fly. Some patterns call for a heavier hackle, and the entire hackle may be used for these flies.

Step 10: When the desired amount of hackle is on, tie the hackle off with two to four thread wraps and apply a half-hitch. Clip off the excess hackle close to the hook shank.

Step 9

Hackle

Wet-Fly Hackle

This exercise begins with the dubbed fur body with tinsel rib from Part Four (page 52). Follow the exercise for the dubbed fur body and tinsel rib, but use a #12 or #14 regular length wet-fly or nymph hook, Monocord or nymph thread, dubbing fur, and tinsel. In addition, you will need hackle from a wet-fly neck, sized to match the hook you use.

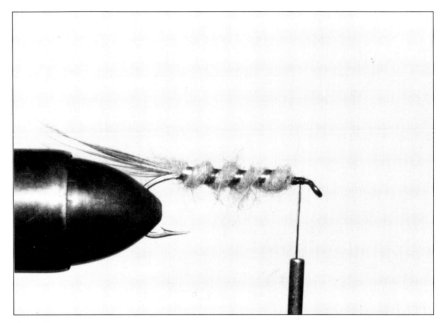

Step 1

Step 1: Tie a dubbed fur body with tinsel rib following the instructions for that exercise on page 52 and using the wet-fly materials listed above. Allow the bobbin to hang.

Step 2

Step 3

Step 2: Using your hackle gauge, select a hackle from the wet-fly neck that is the correct size for the hook you use.

Step 3: Strip the fibers off the bottom quarter of the hackle stem and clip the stem, leaving about an inch of bare quill.

Step 4

Step 5

Step 4: Tie in the hackle stem immediately in front of the body on the side of the shank facing you, at the angle shown in the photograph. Make one loose thread wrap and pull it tight. The concave, or dull, side of the feather should face the hook shank; the shiny side should face you.

Step 5: Make three or four more forward wraps of thread, pulling each snug. Apply a half-hitch and clip the excess stem as close to the shank as you can manage.

WET-FLY HACKLE

Step 6

Step 7

Step 6: A side view of the body with the hackle tied in, clipped, and ready to wind.

Step 7: Grasp the tip of the hackle in your hackle pliers and wind it forward three complete turns. Make each turn tight and abut it smoothly and snugly against the previous wrap. Tie the hackle off with two to four thread wraps and apply a half-hitch.

Hackle

Step 8

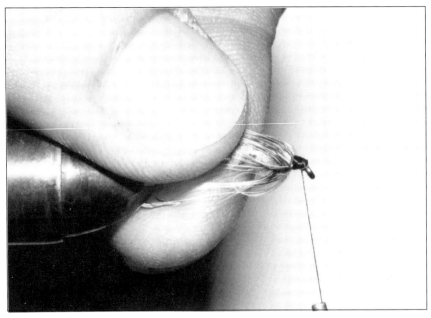

Step 9

Step 8: Clip off the excess hackle close to the hook shank, but be careful not to cut into the wound hackle.

Step 9: Gently slide the thumb and forefinger of your left hand over the eye of the hook and back over the wound hackle, pulling the fibers toward the rear of the hook just far enough to permit a head to be constructed. Pick up the bobbin in your right hand and wrap backward up to twenty-five turns, wrapping over a small portion of the held-back fibers as you do so. These wraps force the hackle fibers back somewhat and form the basis for a smooth tapered head. Let go of the fibers with your left hand when you've completed these wraps.

WET-FLY HACKLE

Step 10

Step 11

Step 10: Continue wrapping to build a smooth tapered head, whip-finish, clip the thread, and apply head cement.

Step 11: A side view of the completed wet fly with tinsel rib and hackle. This is a generalized wet fly that is quite fishable.

Hackle

Palmered Hackle

A number of patterns require that the hackle be spiralled along the hook shank. This is referred to as palmered hackle. For this exercise you will need a #12 or #14 1XL wet-fly or nymph hook, Monocord or nymph thread, chenille, and hackle from a wet-fly neck. In this sequence you will learn how to palmer oversized wet-fly hackle.

Step 1

Step 1: Select an "oversized" hackle from the wet-fly neck. An oversized hackle is one with fibers longer than about one and one-half times the gape of the hook. Strip away all the fibers on the bottom one-third of the hackle stem. Clip the stem, leaving about a half-inch or so with which to tie the hackle on.

PALMERED HACKLE

Step 2: Secure the hook in the vise and tie the thread on just above the point of the hook. Position the hackle stem on the side of the hook that faces you and tie it in, wrapping the thread forward until the stem is in position on top of the hook shank, with the concave, or dull, side of the feather up. Apply a half-hitch and clip the excess stem.

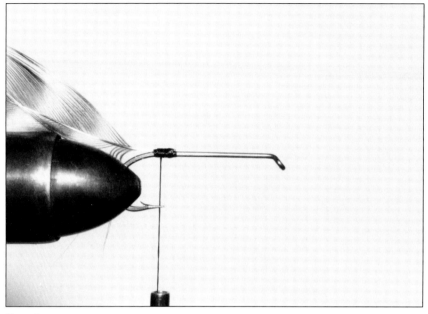

Step 2

Step 3: Construct a chenille body over the tied-in hackle stem, following the instructions for the exercise "Chenille Body," page 58.

Step 3

Hackle

Step 4

Step 5

Step 4: Grasp the tip of the hackle with your hackle pliers and begin wrapping it forward over the chenille body. Leave a gap of about an eighth of an inch between the wraps, as pictured.

Step 5: Continue to wrap forward in a spiral. The final wrap should end on the hook shank just in front of the chenille body. Tie the hackle off at that point with two to four wraps and apply a half-hitch.

Step 6: Clip the excess hackle tip close to the hook shank.

Step 6

Step 7: Wrap a smooth tapered head, whip-finish, clip the thread, and apply head cement. You have tied a Woolly Worm, one of the most famous and effective wet-fly patterns.

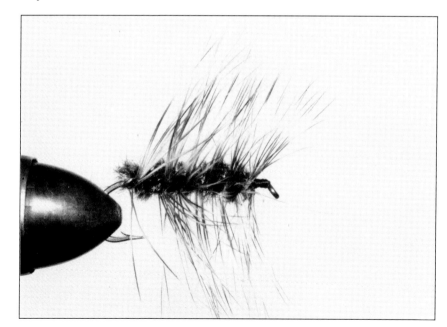
Step 7

Afterword

Times have changed. Fifty years ago the craft of fly tying (or dressing, as some call it) was the domain of relatively few people. Most of them were reluctant to share their knowledge of the subject with the rest of us, which sure made it tough for beginners.

Norm and his brother started out using a large cotter pin held in a standard work-bench vise as their fly-tying vise, along with some snelled hooks, model airplane glue, and their mother's sewing thread and knitting wool. But what to do about hackle? Well, surely Dad wouldn't mind a few bristles missing from his paint brush. After all, it was for a good cause. Later finding a dead Plymouth Rock chicken on the outskirts of town was a real bonanza! Finally, a member of the local rod and gun club and an expert fly tier, James "Hammy" Cunningham, took them under his wing and taught them all he knew about fly tying. Before leaving to serve in the army in World War II, he told his father to give all his fly tying tools and materials to Norm's brother Rich if anything were to happen to him. Hammy was killed in Europe. His vise and other tools are still in use.

Times have indeed changed. The mystique of fly tying no longer exists. Many organizations like Trout Unlimited, Federation of Fly Fishers, and local fish and game clubs have included fly-tying classes, demonstrations, and competitions as an integral part of their programs. Classes on fly tying are even being taught, sometimes for credit, at the high school and university level. There were very few books on fly tying fifty years ago; today we have a wide range to choose from. There are books on how to select materials, how to dye fur and feathers, how to tie certain patterns—you name it. Almost every issue of any popular fly-fishing magazine contains newly devised patterns and a list of materials needed to tie them.

Yes, times have changed, but not all for the better. It wasn't so long ago that we could walk the banks of our favorite trout stream and simply look up at the underside of the overhanging leaves to spot the real-life prototypes of the fly patterns we tied. But the prolific mayfly hatches of years ago no longer exist on many streams. We cannot be sure whether this decline in the mayfly (and other insect) population is due to acid deposition, poisoning of the water through use of herbicides and pesticides, or some other cause. Of this we can be sure—it is happening. Loss of the mayfly is only an indicator. What is the solution? What can we do? The answers to these questions are not simple. But at the least we need to be aware of what is happening, and, using that knowledge, raise the level of consciousness among those who have the power to influence the final outcome.

Appendix

The following is a partial listing of suppliers of fly-tying materials. They all supply catalogs upon request from which you can order various tools and tying supplies. Many of the products from these companies may be found at various fly shops and other retail outlets; their catalogs, however, will give you the full range of their offerings.

Dan Bailey's Fly Shop
P.O. Box 1019
Livingston, Montana 59047

L. L. Bean (request the fly-tying specialty catalog)
95 Main Street
Freeport, Maine 04033

Beckie's Fishing Creek Outfitters
RD 1, Box 310–1
Benton, Pennsylvania 17814

Belvoirdale
P.O. Box 176
Wyncote, Pennsylvania 19095–0176

Fly Fisher's Paradise
P.O. Box 448 Pike Street
Lemont, Pennsylvania 16851

E. Hille Angler's Supply House
815 Railway Street
Williamsport, Pennsylvania 17701

Hook & Hackle Company
P.O. Box 1003
Plattsburg, New York 12901

Jack's Tackle
RD 1, Box 196
Galeton, Pennsylvania 16922

Kaufmann's Streamborn
P.O. Box 23032
Portland, Oregon 97223

Bob Marriott's Fly Fishing Store
2700 West Orangethorpe
Fullerton, California 92633

Orvis (request the fly-fishing and fly-tying specialty catalogs)
Route 7
Manchester, Vermont 05254

Reed Tackle
Box 1250
Marshalls Creek, Pennsylvania 18335

D. H. Thompson
11 North Union Street
Elgin, Illinois 60213

Umpqua Feather Merchants
Box 700
Glide, Oregon 97443